Mysterious
神秘の標本箱
――昆虫――

Cabinet of Insects

丸山宗利
吉田攻一郎
法師人響

KADOKAWA

contents

※昆虫名後のアルファベットは撮影者の頭文字です。
Ⓜ=丸山宗利　Ⓨ=吉田攻一郎　Ⓗ=法師人響
※文章はすべて丸山宗利（P152 コラムは除く）。

興味から慈しみへ

丸山宗利
Munetoshi Maruyama

昆虫が激減している。実に恐ろしいことである。多くの人にとって環境問題は縁遠いものと思われがちだが、昆虫の減少は私たちの身近に迫る重大な問題である。昆虫は世界に500万種以上いるとされ、その形態的、生態的な豊かさを考慮すると、あらゆる生物のなかでもっとも多様な分類群といえる。地球は昆虫の惑星という人もいるほどだ。だから当然、昆虫がいなければ、私たちの世界はまわらない。一例をあげれば、私たちが食べる野菜や果物はもちろん、かなりの植物は昆虫に受粉を依存している。さらに、枯れ木などを分解する役割も大きく、昆虫がいなければ森林は更新されず、いずれ消滅してしまう。その大切さのいっぽうで、昆虫は多くの人に疎まれる存在でもある。多くの子供が昆虫好きな日本人でも、たいていの大人は虫が苦手だ。ベランダに虫が出ても大騒ぎし、殺虫剤を持ち出す人も少なくない。しかし、昆虫の減少が進行している昨今、昆虫への慈しみは環境を守るための原動力となる。たとえば、あちこちにビオトープをつくるなど、身近な自然を保全することがますます重要になっているが、そのときに昆虫好きになれとは言わないものの、昆虫が苦手な人が少なければ、その活動はより有意義な方向に進むだろう。身近な昆虫をむやみに嫌わない。それは大きな第一歩である。どんな物事もそうだが、苦手意識の背景には「よくわからない」ものへの漠然としたおそれがある。昆虫が苦手な人は、昆虫の美しさや面白さを知らない人がほとんどのはずだ。そこで、多くの人に昆虫の魅力を知ってほしい。興味をもつ喜びを知ってほしい。そんな思いで私たちはこの写真集をつくった。昆虫にはあらゆる色があり、色という言葉を超えた彩りがあると言ってもいい。そして昆虫にはあらゆる形や暮らしがある。私たちが想像さえしなかった驚くべき構造や現象もあり、そこには必ず自然の機能美が存在する。本書にはそんなエッセンスを詰め込んだ。先入観を傍らに置いて、とにかく神秘の世界をご覧いただきたい。

（著者を代表して）

Preface

Insects are drastically declining, and this is truly alarming. With over 5 million species, insects are the most diverse group of organisms in form and biology. Without them, our world wouldn't function—many plants, including those we eat, depend on insects for pollination, and insects also play a crucial role in decomposing dead wood, helping forests thrive. Though environmental issues may seem distant to many, the decline in insect populations is a pressing and visible concern. Despite this, insects are often disliked. Even in Japan, where children generally love insects, many adults fear them, sometimes resorting to insecticides at the sight of one. However, in today's dwindling insect populations, fostering even a small appreciation for them can make a difference. While we don't ask that everyone love insects, fewer people disliking them would help conservation efforts. Disliking something often stems from a lack of understanding. Many who fear insects don't know their beauty and fascinating traits. With this photo book, we hope to showcase the incredible diversity of insects—colors beyond words, shapes, behaviors, and natural beauty that will leave you in awe. We invite you to explore this mysterious world.

［凡例］

- 昆虫の種名は、代表的な和名と英訳をのせています。
- 本文下には世界共通の学名を記載しています。「BL」は頭の先端から腹部の先端までの長さ、「Forewing length」は前翅の長さを表します。
- 本文の英訳は、和文の一部を訳したものです。

高精細ギロ

日本のカミキリムシ

Japanese longhorn beetle

最近の研究で多くの昆虫が発音をすることがわかっている。大抵の種は体の一部分を摩擦することによって発音し、ヒトの耳には聞こえないものも少なくない。種内のコミュニケーションに使う場合、ヒトに聞こえる必要がないからである。しかし、カミキリムシに関しては、ヒトの耳にも聞こえる明瞭な音を出す。それは敵に対する威嚇音だからである。ルリボシカミキリは日本を代表する美しいカミキリムシで、青と黒の毛で独特の模様をつくり出している。本種を含む高等なカミキリムシでは、翅の付け根にあるヤスリ状のひだと、前胸の後縁を擦り合わせて音を出す。

鞘翅目／ルリボシカミキリ *Rosalia batesi* (Coleoptera: Cerambycidae: Cerambycinae) (Japan). BL=22 mm.

[撮影メモ] 青い色がすぐに褪せてしまうため新鮮な標本を採集する必要がありました。いざ狙うと難しく、シーズン末期にようやく見つけられた時には思わず声が出ました。(吉田)

Many insects can produce sounds, but some sounds are inaudible to humans, as they are used for communication within the species, not for us to hear. However, certain longhorn beetles make sounds that humans can hear, which are meant to scare off predators. This species is one of Japan's most beautiful longhorn beetles, with striking blue and black patterns. Advanced longhorn beetles like this one produce sound by rubbing file-like ridges near their wing bases against the rear edge of the thorax.

眼ブラシ

カマキリ Mantis

カブキカマキリは西〜中央アフリカに生息する中型の個性的なカマキリで、威嚇するときにカマを左右に開き、内側の派手な模様を目立たせるとともに、翅を広げて裏側の模様を見せる。体も大きく見え、模様のきわどさに天敵となる小鳥などを驚かす効果があるのだろう。また、カマの内側には複眼を掃除するための細かい毛が生えており、さらにそれを拡大すると、汚れをひっかけやすいように、粗い櫛状の形をしているのが面白い。

蟷螂目／カブキカマキリ *Prohierodula picta* (Mantodea: Mantidae: Tenoderinae) (W. & C. Africa). BL=68 mm.

Prohierodula picta is a medium-sized, unique mantis found in West and Central Africa. When threatened, it spreads its fore-legs to reveal bright inner patterns and opens its wings to show the designs underneath, making itself appear larger and more intimidating to predators like birds. Interestingly, the inside of its fore-femour has tiny hairs used to clean its compound eyes, and these hairs are shaped like rough combs to efficiently catch debris, adding to its fascinating adaptations.

ブラウスとスカート

小型のビワハゴロモ Small lanternfly

ビワハゴロモのなかまは日本に在来種がおらず、なじみの薄い昆虫であろう。イネの害虫をふくむウンカや、アオバハゴロモなどに比較的近い。セミのように木の汁を吸うため、世界各地に広まっているある種のビワハゴロモは害虫ともなっている。熱帯では多くの種が繁栄し、いくつかの種は非常に鮮やかな色彩を有する。本種はその名の通り宝石のように美しく、鮮やかな水玉模様の前翅と透き通った青い模様をもつ後翅が印象的で、そのまま洋服のデザインにしたいほどである。この色彩は水分と色素が混じって発色したもので、残念ながら乾燥標本になると美しさは減じてしまう。

Like cicadas, the members of Fulgoridae suck sap from trees, and one Chinese species has become pests worldwide. Many tropical species thrive, with some displaying vibrant colors. This species is jewel-like, with striking front wings adorned with bright polka dots and transparent blue-patterned hind wings. A mix of moisture and pigments creates these colors, but unfortunately, they fade when the insect is dried for preservation.

半翅目／ホウセキヒメテングビワハゴロモ *Saiva gemmata* (Hemiptera: Fulgoridae) (SE Asia). Forewing length=18 mm.

分子間力

オセアニアのタマムシ

Oceanian jewel beetle

熱帯に生息する大型のタマムシの多くは、木の高い場所の葉の上にくらし、それを餌として生活している。タマムシを捕まえたことのある方であればわかるが、ずっしりと重い。それにもかかわらず、木の上を飛び回り葉の上に止まったり飛んだりを器用に繰り返す。そのとき、葉から滑り落ちないか心配になるが、それには及ばない。脚の先にある跗節の裏側に、先端が広がった細い毛がびっしりと生えており、おそらくファンデルワールス力（吸着に関係する分子間力の一種）が関係しているのだろう、それによってしっかりと葉につくことができるのだ。

Many large jewel beetles in the tropics live high in the trees, feeding on leaves. Those who have handled one know how heavy they are, yet they skillfully fly and quickly land on leaves. Despite their weight, they don't slip off the leaves, thanks to the dense, fine hairs on the underside of their tarsi. These hairs likely use van der Waals forces, allowing the beetle to grip the leaf securely.

鞘翅目／オレンジオウサマムカシタマムシ *Calodema mariettae* (Coleoptera: Buprestidae) (Oceania). BL=29 mm.

貴族のガウン

スイコバネガ Eriocraniid moth

小さいながら恐ろしく美しい蛾である。光り輝く鱗粉で織りなされる紫色と金色の模様をもち、まるで中世ヨーロッパの貴族が着ていたガウンのような雰囲気である。幼虫はシラカバなどの葉にもぐって葉肉を食べ、成虫は春先の短い期間に活動する。頭や翅の縁に生える長い毛もすばらしく、小さいからこその造形ともいえるだろう。

鱗翅目／キンマダラスイコバネ *Eriocrania sparrmannella* (Lepidoptera: Eriocraniidae) (Japan). Forewing length=10.2 mm.

This tiny yet stunningly beautiful moth features shimmering purple and gold scales reminiscent of a gown worn by European nobility in the Middle Ages. The larvae feed on the inner layers of birch leaves, and the adults are active briefly in early spring. The long hairs on its head and wing edges add to its delicate beauty, showcasing the intricate design that comes with its small size.

不調和

アフリカのマルカメムシ African globular stink bug

丸い生き物というのには、どこかしら愛嬌がある。ダンゴムシが細長かったら、子供たちの遊び相手にはならなかっただろう。不思議と丸い生き物には穏やかさやかわいらしさを感じるのがヒトであり、もしかしたら健康な赤ん坊にその姿を重ね合わせているのかもしれない。マルカメムシのなかまも一様に丸く、可愛らしい雰囲気のカメムシである。しかしこの種はちょっと違う。オスは丸い胴体に立派な角が生えているのである。おそらくこれを武器にして、メスをめぐって闘うものと思われる。まるで無垢な子供が武器を手にしているようで、ドキッとする姿だと感じてしまう。

半翅目／クワガタマルカメムシ *Ceratocoris cephalicus* (Hemiptera: Plataspidae) (C. Africa). Left: female (BL=16 mm); right: male (BL=27 mm).

Something is endearing about round creatures. Humans often find round shapes gentle and cute, perhaps because they remind us of healthy babies. Members of the Plataspidae are also round and have a charming appearance. However, this particular species is different. The male has a round body but also sports large horns, likely used in battles for mates. The contrast between its innocent, round shape and the fierce horns gives it a surprising and striking look.

衣装に惑う

ニューギニアのゾウムシ

New Guinean weevils

ニューギニアのゾウムシの箱

ホウセキゾウムシはニューギニア島とその周辺の島々に固有のゾウムシのなかまである。大型できわめて硬く、昆虫針を刺すのにも一苦労する。おそらく小鳥などは食べにくいに違いない。派手な色彩はそのことを警告するものであろう。この属は分類に問題があり、明らかに同一の祖先をもつ分類群ではない。地味なゾウムシのなかまから独立して派手な色彩のものが何度も進化しており、それらが一つの属にまとめられている。つまり他人の空似がこのなかに混じっているのである。

The new guinean weevils are native to New Guinea and nearby islands. These large, extremely hard weevils are difficult to pin, making them challenging prey for birds. Their bright colors likely serve as a warning to predators. However, the classification of this group is problematic, as they do not all share a common ancestor. Vivid colors have evolved independently multiple times from normal-colored weevils, and these unrelated species have been grouped together.

鞘翅目／ A specimen box of *Eupholus* spp. (Coleoptera: Curculionidae: Entiminae) (Oceania).

暗がりの幻光

スカシジャノメ Phantom butterfly

中央アメリカから南アメリカの熱帯域には透明な翅をもつチョウがいくつかの分類群（属）にわたって生息している。スカシジャノメのなかまはその代表であり、種によって透明な翅の一部にピンクや紫色、黄色など、鮮やかな色彩をもっている。これらの種は暗い森では影を隠し、止まるとまったく存在がわからなくなる。本種ではその部分が学名のアンドロメダの通り、宇宙を彷彿とさせる深い紫色であり、黒い背景で撮影すると幻光を放ち、その美しさが際立つ。表紙を飾っているベニスカシジャノメ *Cithaerias pireta* は本種の近縁種である。

Several butterfly groups in the tropical regions of Central and South America have transparent wings. The phantom butterflies are a prime example, with some species featuring bright colors like pink, purple, or yellow on their transparent wings. This species, true to its scientific name, andromeda, has deep purple markings that evoke the cosmos. The butterfly on the cover of this book, *Cithaerias pireta*, is a close relative of this species.

鱗翅目／ムラサキスカシジャノメ *Cithaerias andromeda* (Lepidoptera: Nymphalidae: Haeterini) (S. America). Forewing length=48 mm.

四本の顎、四つの眼

南アメリカのクワガタ

South American stag beetle

南アメリカを代表するクワガタの一種で、チリとアルゼンチンのナンキョクブナ林に生息する。オスはクワガタきっての長大な大顎をもち、しかも左右それぞれが基部から二股に分かれている。脚も異様に長く、どこをとっても、一度見たら忘れられない珍奇な形態である。きわめて好戦的なクワガタで、樹液の出る場所をめぐって戦うというが、その際にこれらの形態が有利に働くのであろう。複眼は頭部の側方で分断され、背面と腹面にそれぞれ1対の複眼があるように見える。これにどんな意味があるのかはわかっていない。

This stag beetle, native to Chile and Argentina, is a striking representative of South America. The males have extraordinarily long mandibles, which split into two branches from the base, and unusually long legs, giving them a unique appearance that's hard to forget. The beetle's compound eyes are divided, appearing to have one pair on the top and another on the bottom of its head.

鞘翅目／チリクワガタ Darwin's beetle, *Chiasognathus grantii* (Coleoptera: Lucanidae: Lucaninae) (S. America). BL=76 mm.

17×13=221

周期ゼミ Periodical cicadas

アメリカには3種からなる17年ゼミと、4種から構成される13年ゼミの、合計7種の「周期ゼミ」と呼ばれるセミが生息している。その名の通り、それぞれ17年と13年の長い幼虫期間をもち、別々の発生年をもつ複数の個体群がアメリカの各地で知られている。その進化的な意味はさておき、2024年はイリノイ州を中心に、この周期ゼミが同時に出現し、大発生となった。生息域はほとんど重ならないが、隣接する地域でこれらが同期するのは稀な出来事である。総じて赤い眼に橙色の翅脈をもち、どれもよく似ているが、なかなか美しいセミだと思う。ちなみに大発生のわりにはうるさくなく、日本のクマゼミの集団のほうがはるかにうるさい。

In the United States, there are seven species of "periodical cicadas"-three species of 17-year cicadas and four species of 13-year cicadas. As their names suggest, they have long nymph stages of 17 and 13 years, with different populations emerging in various regions across the country. In 2024, these periodical cicadas emerged simultaneously, primarily in Illinois. They are strikingly similar and quite beautiful with red eyes and orange wing veins.

半翅目／周期ゼミ periodical cicadas (Hemiptera: Cicadidae) (N. America) 17年ゼミの3種
（左頁）17-year cicadas (left page): 1, *Magicicada septendecim*; 2, *M. cassini*; 3, *M. septendecula*; 13年ゼミの4種
（右頁）13-year cicadas (right page): 4, *M. tredecim*; 5, *M. neotredecim*; 6, *M. tredecassini*; 7, *M. tredecula*.

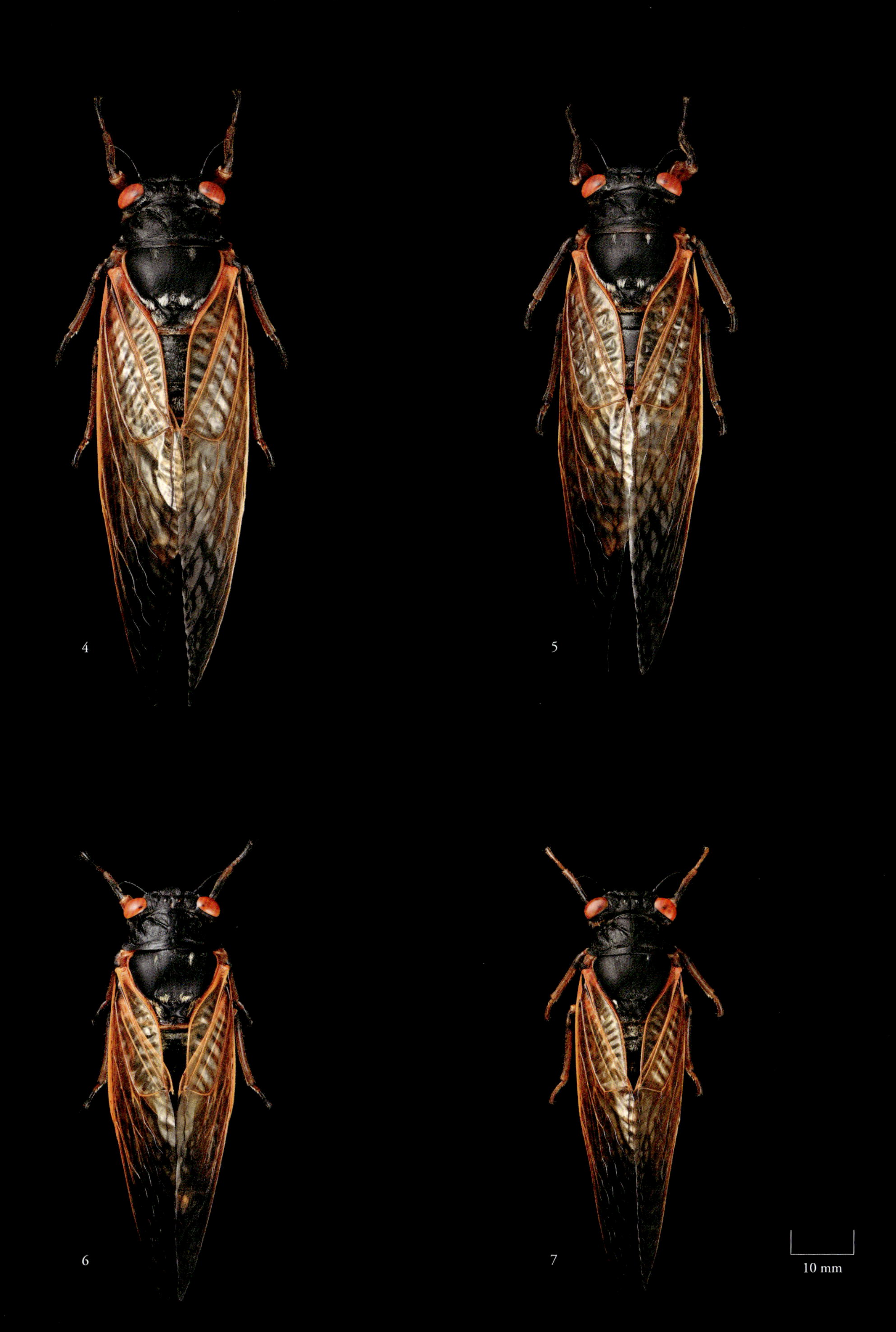

［撮影メモ］見た目に差が無いため、写真が混ざらないよう管理に気を使いました。（法師人）

1

5

6

［撮影メモ］生きた女王を傷つけないように撮影するのに苦労しました。（丸山）

神農

シロアリ Termites

キノコシロアリのなかまは、その名の通りキノコを育てるシロアリで、農業をするシロアリともいえる。枯れ草や朽ち木をかじって飲み込み、出した「偽糞」をかためて菌園（畑）をつくり、そこに菌糸を植え付けて、それが増えると収穫して食べ物にする。枯死した植物そのものを食べるより、キノコのほうが栄養が豊富なのである。菌園は地下にあるが、ときに大きなキノコが地面を突き破って地上に現れる。それは非常においしく、ヒトにとっての高級食材ともなる。女王は巨大で、1日に数千個の卵を産むという。シロアリには、卵、幼虫、働きアリ、兵アリ、王、女王のカーストがあるが、ここにはすべてを掲載している（卵は女王の腹部に付着している）。

As their name suggests, fungus-growing termites cultivate fungi, making them the "farmers" of the termite world. They chew on dead grass and wood, produce a material called "pseudo-feces," and use it to create fungal gardens. These fungi provide more nutrition than the dead plants themselves. The queen is enormous and can lay thousands of eggs daily. Termites have various castes, as shown here.

ゴキブリ目／オオキノコシロアリの一種 *Macrotermes annandalei* (Blattodea: Isoptera: Termitidae: Macrotermitinae) (SE Asia). 1, 女王 (queen); 2, 王 (king); 3, 幼虫 (nymph); 4, 働きアリ (worker); 5, 大型兵アリ (large soldier); 6, 小型兵アリ (small soldier).

背中のファスナー

ツツハムシ

Case-bearing leaf beetle

イモムシと同じように植物の葉を食べるが、下にある和名の通り自分はイモムシの糞の形をしているという不思議な昆虫である。その体の精緻な収納構造は見事で、敵におそわれたときなどには、触角と脚をすべて溝に収納して、まるで凄腕の職人がつくった工芸品のように、シンプルな円筒形になる。そして前翅の形も面白く、ギザギザとした会合部で、きっちりと閉じるようにできている。これでしっかりと飛ぶこともできるというのもすごいことだ。

This beetle feeds on plant leaves, like caterpillars, but resembles caterpillar droppings. Its body has an impressive storage structure, allowing it to tuck its antennae and legs into grooves when threatened, forming a simple cylindrical shape that looks like a master craftsman's work. Its forewings are also fascinating, with serrated edges that close tightly. Despite this compact design, it can still fly efficiently, which is remarkable.

鞘翅目／ムシクソハムシ *Chlamisus spilotus* (Coleoptera: Chrysomelidae: Cryptocephalinae) (Japan). BL=3.1 mm.

樹の皮の下で

ヒラタムシ Flat bark beetles

ヒラタムシはその名の通り平べったい甲虫の一群である。日本には３種の大型種が生息し、いずれも絹のような光沢からなる美しい色彩をもっている。幼虫も平たい体形で、枯れた木の樹皮下に生息し、他の昆虫を捕食する習性をもつ。ヒラタムシを含むヒラタムシ下目はさまざまな環境に多様化した一群で、テントウムシやゾウムシ、ハムシなども含む。

The flat bark beetle genus *Cucujus* is a group of beetles with flat bodies. In Japan, there are three large species, each displaying a subtle, beautiful sheen. Their larvae are also flat and live under the bark of dead trees, where they prey on other insects. The suborder Cucujiformia that includes flat bark beetles has diversified into various environments and includes other well-known groups like ladybirds, weevils, and leaf beetles.

鞘翅目／ヒラタムシのなかま (Coleoptera: Cucujidae) (Japan). **1, ルリヒラタムシ** *Cucujus mniszechi*; 2, **ベニヒラタムシ** *C. coccinatus*; 3, **エゾベニヒラタムシ** *C. opacus*.

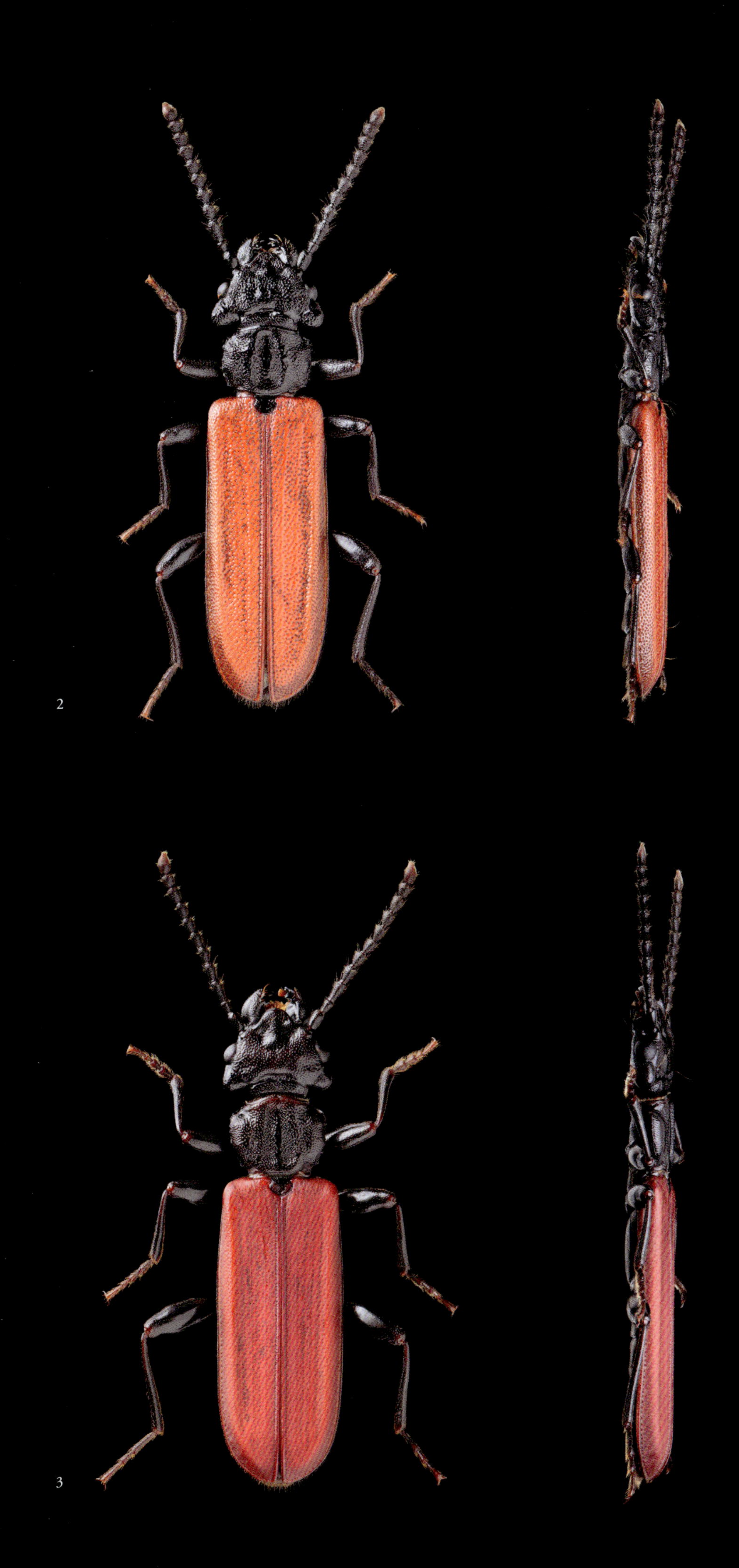
2
3

生き返る標本

クサビウンカ Japanese issid planthopper

10年ほど前、昆虫の後脚の付け根（転節）に歯車のような構造が見つかり、有名科学誌「Science」に掲載され、話題となった。その虫はヨーロッパ産のクサビウンカの一種の幼虫である。後脚でジャンプをする際、左右の脚の動きを完全に同調させるため、その歯車構造があるという驚きの機能である。日本にも同属種がいるので、日本産種はどうなっているのかが気になっていた。そして愛媛大学ミュージアムの収蔵庫に古い標本を見つけた。顕微鏡でつぶさに観察すると、予想は的中し、日本産種の幼虫にもたしかにその構造があることがわかった。最初の発見からは出遅れたが、古い標本の意味を改めて感じることになった。

About ten years ago, a gear-like structure was discovered at the base of an insect's hind legs (trochanter), making headlines in the journal *Science*. The insect was a nymph of a European issid planthopper species. This gear helps synchronize the movements of the hind legs during jumping. Curious about whether the same structure existed in Japanese species of the same genus, we examined a specimen from an old collection. Our hunch was correct—the same structure was found in the Japanese species.

半翅目／カタビロクサビウンカ *Issus harimensis* (Hemiptera: Issidae) (Japan). BL=7 mm (adult); 4 mm (nymph).

［撮影メモ］拡大写真の中央の少し下に、横からの歯車構造が見えると思います。（丸山）

格納構造

コブゴミムシダマシ Ironclad beetle

中央アメリカに繁栄するコブゴミムシダマシのなかまで、大型かつ非常に堅牢な甲虫である。そんな硬い甲虫でも弱い部分がある。それは触角と脚である。触角は重要な感覚器であり、傷つけられてしまえば生活に支障がある。枯れ木上に生活し、外敵に見つかったときには脚をちぢめて転がり落ちるが、その際に触角を一番大切な宝物のように、前胸の腹面にある溝にしまい込む。その溝は職人があつらえたかのように触角がぴったりと収まるようにできている。

鞘翅目／フチナミマダラコブゴミムシダマシ *Zopherus jourdani* (Coleoptera: Zopheridae: Zopherinae) (C. America). BL=30 mm.

This beetle from Central America has a large and rigid body. Despite its hard body, its antennae and legs are vulnerable. The antennae are vital sensory organs, and damage to them would disrupt its life. When threatened, it curls up and falls from dead wood, tucking its precious antennae into grooves on the underside of its thorax. These grooves fit the antennae perfectly as if explicitly crafted for protection.

ねじれた生活

ネジレバネ Twisted-wing insect

ネジレバネという名前を聞いたことがあるだろうか。撚翅目（ネジレバネ目）という目に属し、すべてが寄生性種からなる。メスは翅がなく、多くの種ではウジ状で、寄主の体のなかに埋没し、頭部周辺だけを出して生活している。オスはある時期になると寄主の体から飛び出し、数時間の短い寿命の間、高速で飛び回り、メスを探して交尾を行う。写真はクロスズメバチに寄生する種のオスで、巨大な触角と、大きな複眼がよくわかる。頭部のうしろには小さな棒状の前翅が見えるが、これが捻じれているのが名前の由来である。この前翅は感覚器として飛行のバランスをとるのに使うとされている。メスはハチの腹部に寄生するが、寄生率は高くない。そのメスを短時間で探し出す途方もない冒険にこれらの道具が必要なのだろう。

This insect belongs to the order Strepsiptera, and they are entirely parasitic. Females are wingless, often maggot-like, and live inside their host's body with only the head exposed. Conversely, males emerge from their host for a brief adult period, flying rapidly for just a few hours to find a female and mate. The photo shows a male of a species parasitizing a hornet species, with large antennae and compound eyes that help locate a female in the hornet's abdomen.

撚翅目／ヤマトネジレバネ *Xenos vespularum*, male (Strepsiptera: Xenidae) (Japan). BL=5 mm.

［撮影メモ］液浸標本の表面が乾いた少しの間が撮影のチャンス。時間との戦いでした。（吉田）

月の蛾

ヤママユガ

Sturniid moth

日本のオオミズアオやオナガミズアオと近縁でよく似ているが、より強壮で長い尾状突起をもつ。この尾状突起はヤママユガのなかまに広く見られ、最近の研究では、捕食者であるコウモリが狩りに使う超音波をずらし、体の中心部を狙われないようにする効果があることがわかった。それにしても美しい青緑の色彩は新鮮な本種の標本ならではの色である。英名を「luna moth（月の蛾）」というが、きっとこの蛾が月夜に照らされたら幻想的だろう。

The tail-like appendages found in many silkmoths of the family Saturniidae, including this species, are known to disrupt bats' echolocation, preventing them from targeting the moth's body. The beautiful blue-green coloration is unique to fresh specimens of this species. Its English name, the "luna moth," suggests a magical appearance when illuminated by moonlight.

鱗翅目／アメリカオオミズアオ Luna moth, *Actias luna* (Lepidoptera: Saturniidae) (N. America). Left: male (Forewing length=55 mm); right: female (Forewing length=53 mm).

予想外

チビゴミムシ Trechine ground beetle

地下や洞窟などの暗黒の世界にもさまざまな昆虫が生息するが、チビゴミムシはその代表的な一群である。どれもオサムシをそのまま小さくし、飴色にしたような甲虫である。移動能力が低く、隔絶された環境に生息するため、地域ごとに多くの種に分化しており、日本だけで400種以上が知られている。ただ、日本で複眼が退化した種が見つかるのは北海道から九州までの温帯の島だけで、多くの調査の結果、琉球列島にはいないものと長らく考えられてきた。しかし、2021年に衝撃が走った。沖縄島で、日本で最も洞窟へ適応した種が発見されたのである。しかも非常に大型で見事なチビゴミムシだった。

Many trechine ground beetles live in dark environments like underground or caves. Isolated in different regions, they have speciated into many species, with around 400 species known in Japan alone. After extensive surveys, they were thought to be absent from the Ryukyu Islands, but in 2021, a remarkable discovery was made. A highly cave-adapted, large species of trechine ground beetle was found on Okinawa Island, stunning researchers with its impressive shape and size.

鞘翅目／オキナワアシナガメクラチビゴミムシ *Ryukyuaphaenops pulcherrimus* (Coleoptera: Carabidae) (Japan). BL=5.5 mm.

地上絵

南アメリカのゴミムシダマシ

South American darkling beetles

砂漠のような環境を好む昆虫にゴミムシダマシ科の甲虫がいる。アフリカのナミブ砂漠にも多数の種がいるが、南アメリカのアンデス山脈周辺の乾燥地にもゴミムシダマシが大繁栄しており、とくに Nycteliini 族の多様性が素晴らしい。面白い模様のものが多く、写真の種はナスカの地上絵のような、細かい毛からなる独特な放射状の模様が特徴的である。その意味はまったくもって不明だが、地表で体の輪郭をぼかす効果があるのだろう。

The darkling beetle family Tenebrionidae includes species that thrive in desert-like environments. Many species live in South America's dry regions around the Andes Mountains. Many have fascinating patterns, and the species shown here features unique radiating designs made of fine hairs resembling the Nazca Lines. Though the purpose of these patterns is unclear, they likely help blur the beetle's outline against the ground.

鞘翅目／南アメリカのゴミムシダマシ *Gyriosomus* spp. (Coleoptera: Tenebrionidae: Pimeliinae: Nycteliini) (S. America). BL=ca. 20 mm.

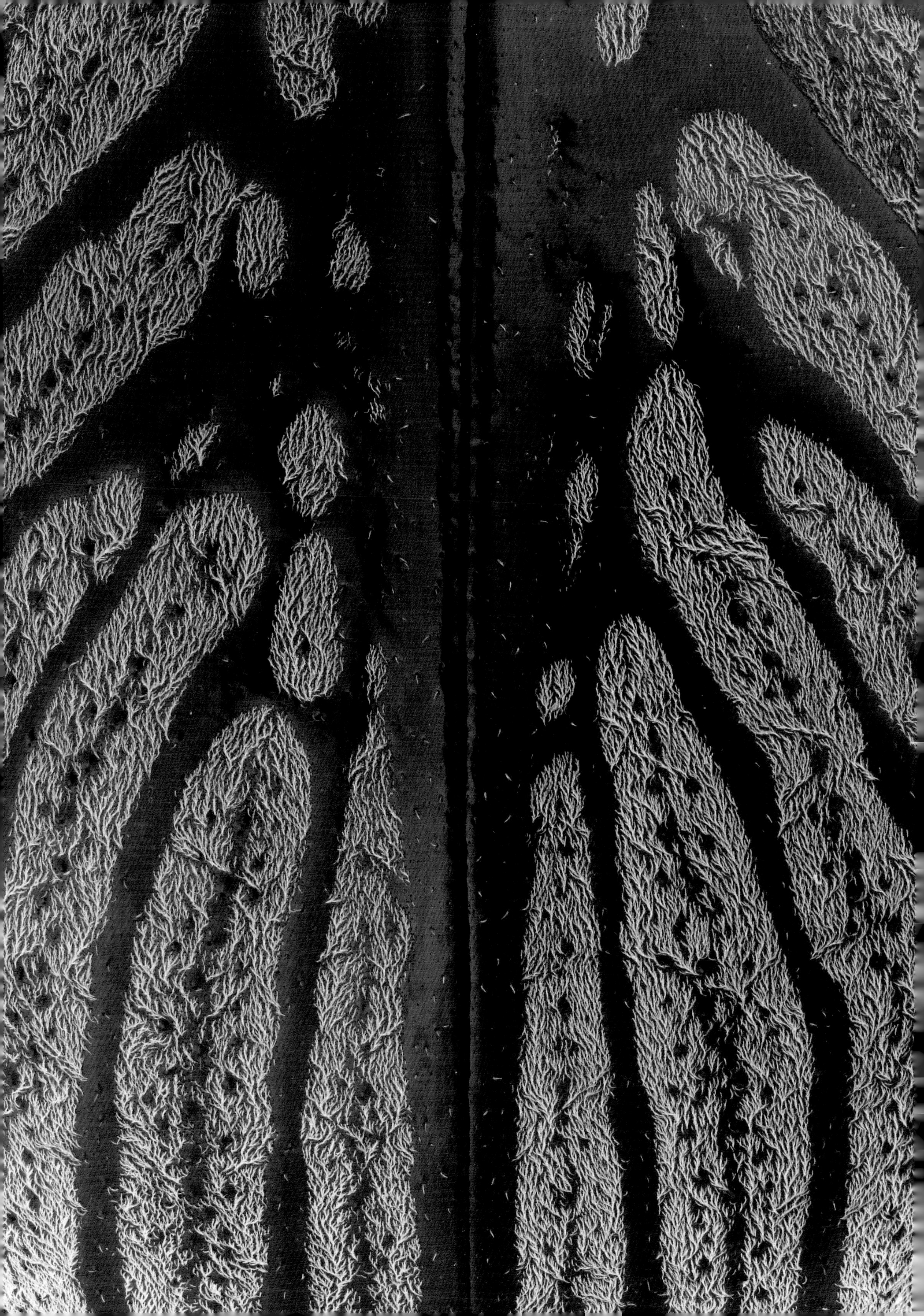

南アメリカの
ゴミムシダマシの箱

アンデス山脈の乾燥地帯に生息するゴミムシダマシである。一見黒い虫の詰まった箱に見えるかもしれないが、どれも個性的な姿をしている。箱の下のほうに、北アメリカ産の数種も含む。

These darkling beetles live in the dry regions of the Andes Mountains. At first glance, the box may appear filled with plain black insects, but each has a unique appearance. The lower part of the box also includes species from North America.

鮮烈目くらまし

バッタ Grasshopper

南アメリカに生息する大型のバッタで、多くの大型のバッタのように草原には生息せず、高木からなる熱帯林に生息している。高い木の幹にとまっていることが過去に目撃されているが、普通はたまに灯火に飛来する程度で、それ以上の生態はわかっていない。後翅は目の覚めるような美しい紫色で、飛んで逃げる際に目立つことで、着地後の姿に目が行かないようにする効果をもつのかもしれない。熱帯の生物多様性には未知なことが多いが、大型の昆虫の生態に関しても不明なことばかりである。

直翅目／シタムラサキオオバッタ *Titanacris albipes* (Orthoptera: Romaleidae: Romaleinae) (S. America). BL=60 mm.

This large grasshopper species from South America lives not in grasslands, as many large grasshoppers do, but in tropical forests with tall trees. It has been seen resting on tree trunks, and occasionally flies to lights at night, but little else is known about its behavior. Its hind wings are a striking purple, possibly used to startle predators when it flees. While much remains unknown about tropical biodiversity, the ecology of even large insects like this is still largely a mystery.

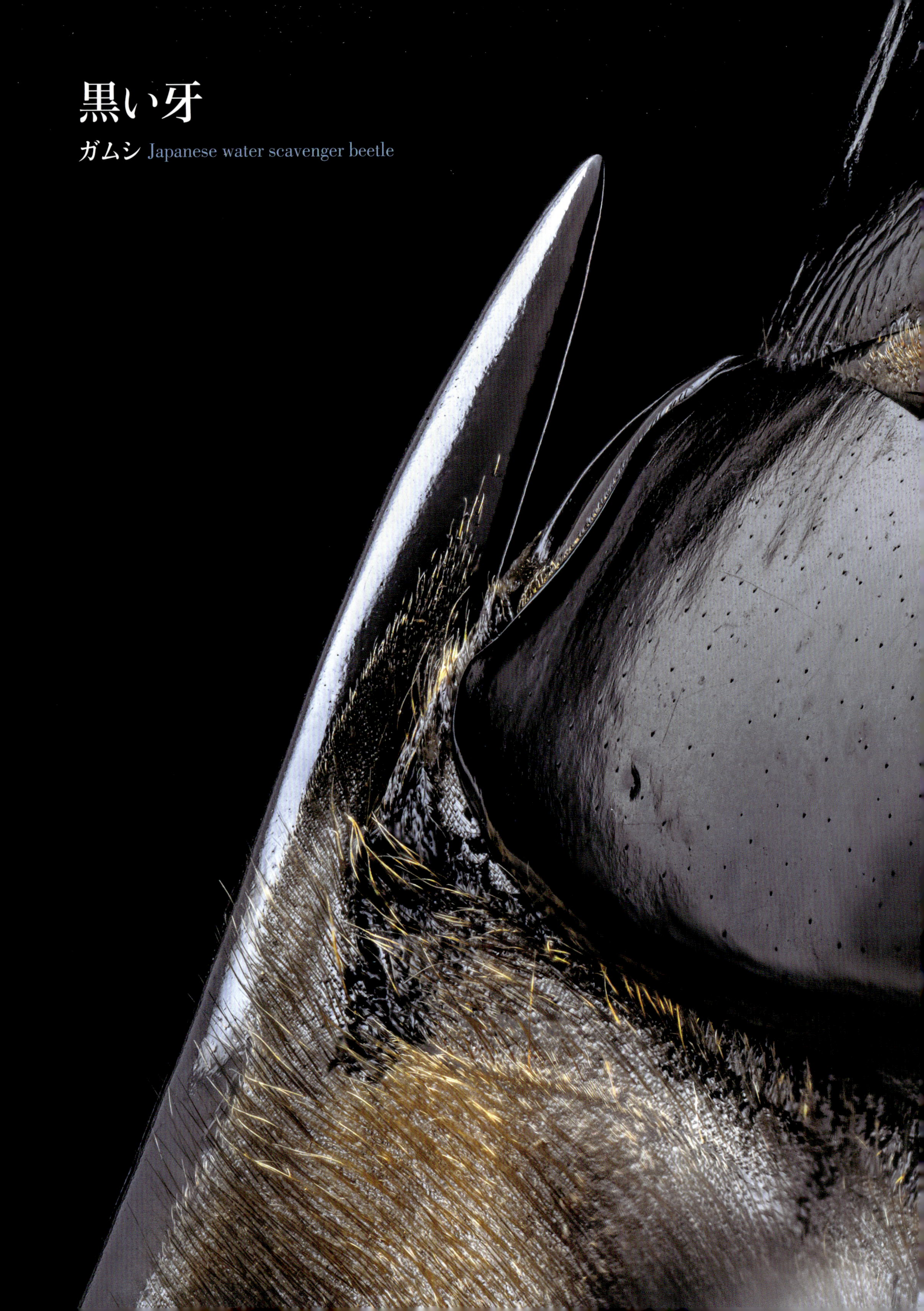

黒い牙

ガムシ Japanese water scavenger beetle

ガムシを漢字で「牙虫」と書く。胸部の腹面に鋭いトゲのようなものがあり、それを牙に見立てたとされる。ゲンゴロウに似ているが、まったく違うなかまである。ゲンゴロウはオサムシなどに近く、ガムシはコガネムシなどに近い。別々に進化して水中で生活するようになった甲虫なのである。ただ、ガムシはゲンゴロウにくらべて不器用に泳ぐ。それはさておき、美しい流線形と見事なトゲ。なかなか渋い美しさがある。

鞘翅目／ガムシ *Hydrophilus acuminatus* (Coleoptera: Hydrophilidae) (Japan). BL=36 mm.

This water scavenger beetle has sharp, fang-like spines on the underside of its thorax. Although it resembles the diving beetle, they belong to completely different groups. Diving beetles are related to ground beetles, while water scavenger beetles are closer to scarab beetles. Both evolved independently for life in water. Compared to diving beetles, water scavenger beetles are clumsier swimmers.

恋の道具

ジョウカイモドキ Soft-winged flower beetle

昆虫の触角は、ヒトでいえば、主に鼻のような役割をもつ。同時に、舌、耳、指先、ときに目の役割さえ担うこともある。つまり、五感を司る重要な感覚器である。さらに別の役割もある。甲虫などでは、触角に分泌腺や把握の機能を備え、それを配偶行動に使うことがある。ジョウカイモドキは小型で体のやわらかい甲虫である。途中の節がふくらんだ、変わった形の触角をもつものがおり、種ごとにその形も異なる。機能は不明だが、オスだけがもつことを考えるとおそらく配偶行動に使われるのであろう。よく見ると、その部分は立体的で美しい形をしている。

Insects' antennae are crucial sensory organs and can also have additional roles. For example, some beetles use them in mating behaviors, with special functions like secretion or grasping. The males of the soft-winged flower beetles have unusual antennae with bulging segments varying by species. While their exact function is unknown, since only males have them, they are likely used in mating.

鞘翅目／イソジョウカイモドキ *Laius asahinai* (Coleoptera: Melyridae: Malachiinae) (Japan). BL=4.3 mm.

舶来キャンディ

ヒゲブトハナムグリ Bumble bee scarab beetle

3月になると八重山諸島に少なからぬ採集者が訪れる。お目当ては珍種のカミキリムシ数種と、このコガネムシである。採り方にはいくつかあるが、木の上を飛び回るオスを網で捕まえるのが一般的である。運が良ければ、ヤンバルアワブキの花に集まるものを採集することもできる。本種の特徴は、なんといっても美しい金属光沢とその色彩の豊かさで、「金色がとれた」「青がとれた」と、何匹採っても興奮は冷めやらない。このように集めてみると色とりどりの銀紙に包まれた飴玉のようでもあり、採集したときの甘美な気持ちを思い出す。一見乱獲のように思えてしまうかもしれないが、森林地帯に広く生息し、しかも手が届く一部の場所でしか採集できないため、森がある限り減るようなことはない。

In March, many collectors visit Yaeyama Islands of Okinawa to search for this particular scarab beetle famous for its stunning metallic sheen and vibrant colors. While this might seem like over-collecting, the beetles widely spread across forest areas, and since they can only be caught in inaccessible places, their numbers are likely to stay the same as long as the forests remain intact.

鞘翅目／オオヒゲブトハナムグリ *Amphicoma splendens* (Coleoptera: Glaphyridae: Amphicominae) (Japan). BL=ca. 15 mm.

[撮影メモ] 現地でお会いした数名の方々に数日分の採集品をお借りしました。(丸山)

きらら

シミ
Silverfishes

［撮影メモ］鱗毛が非常に取れやすく、麻酔後に動かすのに苦労しました。しかしそのキラキラも美しかったので、撮影してみました。（丸山）

シミは漢字で「紙魚」と書き、その名の通り、全身が鱗のような毛で覆われている。陽が当たるとその鱗がきらめくことから、「きららむし」とも呼ばれた。いくつかの種は家屋性で、書物や壁の隙間を魚が泳ぐようにすばやく走り回り、さまざまな有機物を食料とする。ときに本や布をかじり、図書や衣類の害虫とされることもあるが、現代では気密性の高い家屋が多くなり、人家で見られることは稀となっている。きわめて原始的な昆虫で、成虫も翅をもたず、成虫になってからも脱皮して成長するなど、昆虫の祖先的な特徴を多く残している。ここに示したヤマトシミ（左）とセイヨウシミ（右）は身近に見られたシミの代表である。その名から想像するように、前者が在来種、後者が外来種とされるのが一般的だが、その実よくわかっていない。

The silverfish, named for its scale-like, silvery body covering, includes some species that live indoors. They dart quickly along books and wall crevices, feeding on various organic materials and occasionally classified as pests. However, due to modern airtight homes, they are now rarely seen indoors. These insects retain many ancestral traits, such as lacking wings and continuing to molt even as adults.

総尾目／シミのなかま *Lepisma saccharina* (Zygentoma: Lepismatidae) (Japan). ヤマトシミ *Ctenolepisma villosa* (left); セイヨウシミ (right).

戦の面

ドロバチ Potter wasp

立体的な形をした見事な大顎、アフリカらしい色彩、実に美しいハチである。オスのみが発達した大顎をもつ。メスは泥を集めて細長い壺のような巣をつくりながら、そこに産卵する。そして幼虫が孵化すると、イモムシを噛み砕いて与え、充分に大きくなると泥で蓋をして、世話をやめる。幼虫はそのなかで蛹となり、羽化すると泥の巣をやぶって外に出る。メスの巣の場合、出てきたメスとすぐに交尾をするため、巣の外でオスが待ち構える。そして、そこでオス同士が鉢合わせになると、メスを待つ権利をめぐって戦うのだが、そのときにこの大顎が役に立つ。

膜翅目／オオキバドロバチ *Synagris cornuta* (Hymenoptera: Vespidae) (Cameroon). BL=28 mm.

This African wasp is stunning with its impressive mandibles and vibrant colors. The female builds a long, vase-like nest from mud, where she lays her eggs. After the larvae hatch, she feeds them by chewing up caterpillars, and once they are large enough, she seals the nest with mud. The larvae pupate inside and, upon emerging, break out of the nest. Males wait outside the female's nest to mate with the newly emerged females. If two males encounter each other, they fight for the right to mate, using their large mandibles in the battle.

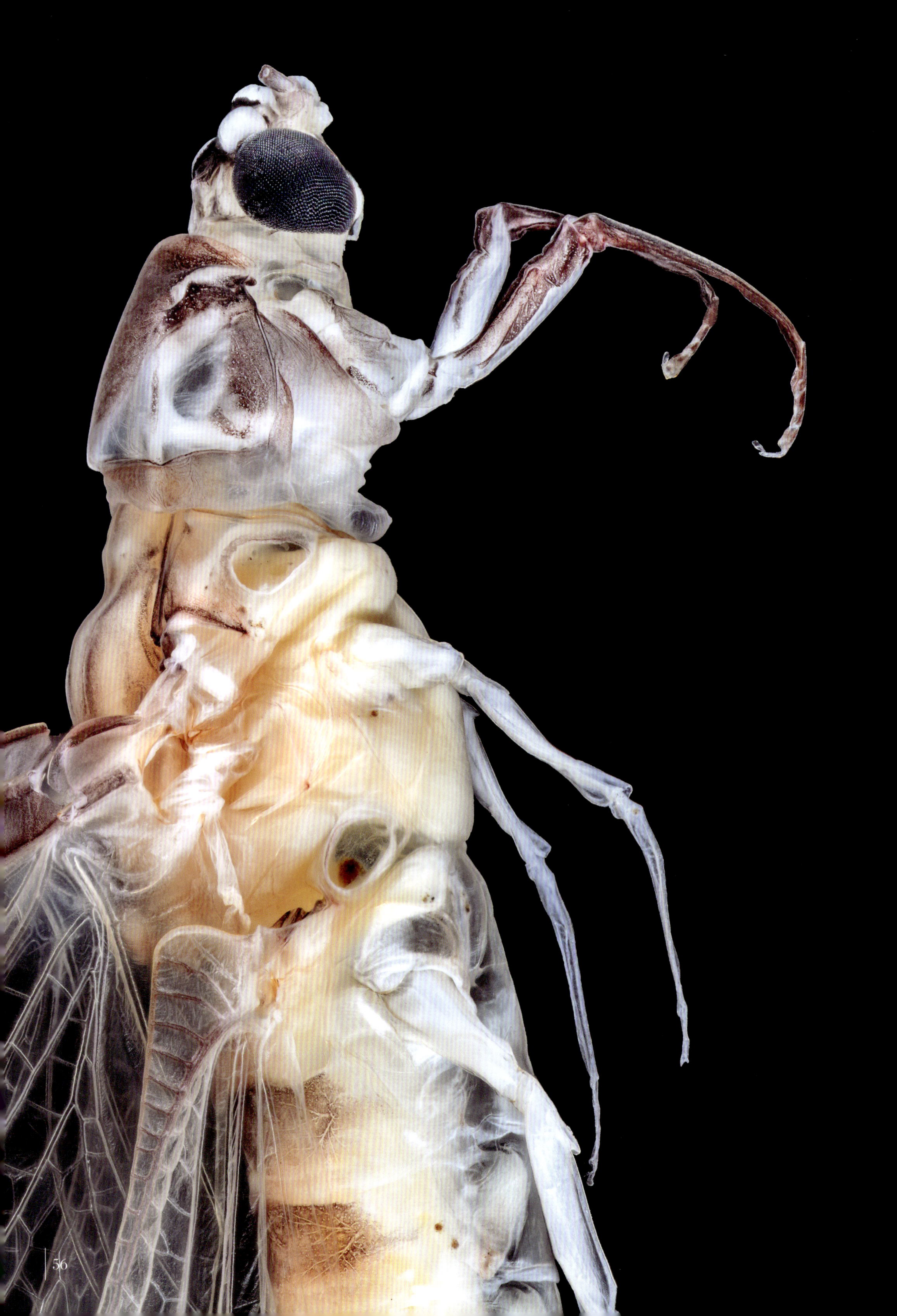

精白

カゲロウ Mayfly

カゲロウの成虫は短命で、交尾と産卵だけに特化したきわめて節約的な姿をしている。オオシロカゲロウのなかまはその最たるもので、水面から羽化した成虫は数十分だけ水面を飛び交い、個体群によっては交尾をせずに産卵し、その短い成虫期を終える。どこかに止まって休むことは計算されていないので、何かにつかまる脚は退化して機能を失い、着水と同時に産卵する。外骨格さえも薄く節約され、腹部の大半を占める卵にその分の栄養が投資されている。本種の研ぎ澄まされた形態は進化の見事な産物だが、その儚さについつい切ない気持ちになってしまう。

蜉蝣目／オオシロカゲロウ *Ephoron shigae* (Ephemeroptera: Polymitarcyidae) (Japan). BL=19 mm.

[撮影メモ] 一晩限りの大量発生を見つけ出すのが大変でした。(法師人)

Adult mayflies have a short lifespan, designed specifically for mating and laying eggs. The species is an extreme example of this efficiency. After emerging from the water, adults only fly for a few minutes, and some populations even skip mating before laying eggs, quickly ending their brief adult phase. They don't rest, so their legs are reduced, and they lay eggs as soon as they touch the water. Even their exoskeleton is thin, with most of their energy invested in the eggs that fill their abdomen.

美意識の西と東

ゴミムシ Ground beetles

ナガゴミムシは主に北半球の温帯域に1000種以上が生息するオサムシ科の大きな一群である。日本にも多数の種が生息し、湿地から山地の石の下まで、種によって生息環境はさまざまである。多くの種は真っ黒く、それはそれで渋い美しさがある。いっぽう、ヨーロッパに行くと、黒い種も多いが、赤や緑色の派手な金属光沢を有する種も少なくない。近い分類群でこのように色彩の傾向が大きく異なるのは、生物学的に興味深い課題である。左頁にヨーロッパ産種、右頁に日本産種を示した。

鞘翅目／ナガゴミムシのなかま *Pterostichus* spp. (Coleoptera: Carabidae: Harpalinae). 1, *Pterostichus pilosus*; 2, *P. burmeisteri*; 3, *P. externepunctatus*; 4, *P. macrogenys*; 5, *P. thunbergi*; 6, *P. prolongatus* (1-3: Europe; 4-6: Japan).

The ground beetle genus Pterostichus consists of over 1,000 species in mainly temperate regions of the northern hemisphere. Many species are found in Japan, with habitats ranging from wetlands to under-stones in mountainous areas. Most are jet black, possessing a subtle, elegant beauty. In contrast, Europe has not only black species but also brightly colored ones with metallic red or green hues. The variation in coloration within such closely related groups is an intriguing biological puzzle.

4
5
6
5 mm

1

2

3

遙かなる影

コオロギとゾウムシ Cricket and weevils

フィリピン諸島にはカタゾウムシという非常に硬いがために捕食者に嫌われるゾウムシが繁栄している。そしてさまざまな昆虫がそれに擬態し、捕食の難を逃れようとする。このコオロギはカタゾウムシに擬態するとされる種で、美しい斑紋に加え、金属光沢のある前胸や脚が甲虫らしさを強調している。カタゾウムシの色彩はさまざまだが、黄色や白の紋をもつ種が多い。本種は特定のカタゾウムシというわけではなく、黄色や白の紋のものに漠然と擬態しているものと思われる。一般にモデルより擬態者のほうが数が圧倒的に少ない。本種も非常に稀な種で、写真が図示されるのは初めてであろう。

直翅目／1, カタゾウムシ擬態コオロギ *Scepastus pachyrrhynchoides* (Orthoptera: Oecanthidae: Podoscirtinae) (BL=12.5 mm); 2, クロライトカタゾウムシ *Pachyrhynchus chlorites* (Coleoptera: Curculionidae) (BL=16 mm); 3, アカアシカタゾウムシ属の一種 *Metapocyrtus* sp. (ditto) (BL=13.6 mm) (SE Asia).

Members of the jewel weevil tribe Pachyrhynchini possess rigid bodies, and predators tend to avoid them. Many insects mimic these weevils to avoid predation. This cricket species is thought to mimic Pachyrhynchini, with its metallic sheen on the pronotum and legs adding to its beetle-like appearance. While the weevils come in various colors, many have yellow or white markings. This cricket does not mimic a specific weevil species but somewhat resembles weevils with such markings in general.

原初の鉗子

シリアゲムシ Scorpionfly

シリアゲムシは原始的な完全変態昆虫で、多くの完全変態昆虫の幼虫がわずかな単眼しかもたないのに、シリアゲムシの幼虫は複眼のような眼をもっている。成虫も原始的な姿で、その特徴の一つに上下でほぼ同じ形の翅がある。このムカシシリアゲムシは丸く大きな翅をもち、たたむと平たくなることで、他のシリアゲムシとはかなり異なる。しかし、細長い頭部はシリアゲムシ全体に共通する特徴である。一見して数億年前の太古の時代を彷彿とさせる原始的な姿の昆虫である。突起の多い特徴的な交尾器にも注目されたい。このような部分も淘汰をあまり受けていない原始的な姿に思える。

The scorpionflies are among the holometabolous insects. This species is particularly primitive, having large, round wings that fold flat, setting it apart from other scorpionflies. However, its elongated head is a common trait across the group. Its appearance evokes a sense of ancient times, hundreds of millions of years ago.

長翅目／ムカシシリアゲムシ *Notiothauma reedi* (Mecoptera: Eomeropidae) (S. America) . Forewing length=47.5 mm.

異形の煙突

ツノゼミモドキ Aetalionid treehopper

ツノゼミ科に近い昆虫にツノゼミモドキ科という一群がある。大部分の種は南アメリカに生息するが、1属2種だけはアジアに生息している。そのひとつがこの種である。腹部末端に煙突のような突起があり、まさに奇虫である。その不思議な形態から多くの学者がその系統的な位置について頭を悩ませてきた。ツノゼミモドキ科としては世界最大で、複雑な網目模様をもつ翅脈も特徴的である。62頁で示したムカシシリアゲムシも同じだが、このように不規則かつ複雑な翅脈をもつものは、たいてい原始的な昆虫であり、本種もそのような昆虫とみなしてよいだろう。

This species has a strange, chimney-like projection at the end of its abdomen, making it truly unique. Its unusual form has puzzled many scientists regarding its classification. It is the largest species in the family Aetalionidae and is known for its complex wing venation with intricate mesh patterns. Like the ancient scorpionfly shown on another page, insects with such irregular and intricate wing veins are typically considered primitive.

半翅目／ツノゼミモドキ *Darthula hardwickii* (Hemiptera: Membracoidea: Aetalionidae) (E. Asia). BL=27 mm.

網底の流線形

ハナノミ

Tumbling flower beetle

ハナノミは尻すぼみの流線形という不思議な形態で、美しい紋をもつものも多く、小さいながらも魅力ある甲虫である。そんなハナノミを捕虫網で採集した時、誰でも経験する歯がゆい思いがある。網の底でハナノミがピョンピョンと跳ねて、なかなかつまめない上に、やっとつまんだと思っても、指の間からスルリと抜けてしまうのである。流線形の体形や、長い微毛がこのときに役立っていると考えられる。おそらく捕食者にとっても厄介な昆虫だろう。美しい紋は「捕まえても苦労するだけ無駄だよ」ということを示す警告色かもしれない。漢字で書くと「花蚤」だが、ノミもまったく同様の行動を示すので、体を表した名前だと思う。

鞘翅目／ウスキボシハナノミ *Hoshihananomia kurosai* (Coleoptera: Tenebrionoidea: Mordellidae) (Japan). BL=9 mm.

Tumbling flower beetles have a unique, spindle-shaped body that narrows at the end, and many species feature beautiful patterns, making them small but charming insects. However, anyone who has tried to catch them with a net knows the frustration—they hop around and are difficult to hold, and even when caught, they often slip away between your fingers. This behavior likely makes them tricky prey for predators as well. Their vibrant patterns might serve as a warning, signaling that catching them isn't worth the effort.

池畔の紫水晶

日本のトンボ Japanese dragonfly

チョウトンボは水草が豊富で明るい池を飛ぶトンボで、その名の通り、翅に美しい模様をもち、チョウのようにひらひらと優美に舞う美しいトンボである。その模様は金属光沢のある濃い紫色であり、他のどの昆虫にも見られない独特な色彩をつくっている。また翅脈の凹凸が複雑に反射し、夏の池にきらめく紫水晶のようである。

蜻蛉目／チョウトンボ *Rhyothemis fuliginosa* (Odonata: Libellulidae) (Japan). BL=34 mm.

［撮影メモ］記憶色と実物の色彩のギャップが生まれないよう、ライティングに気をつけました。（法師人）

This dragonfly is found in bright ponds rich in aquatic plants. It gracefully flutters like a butterfly with beautiful wings adorned in a metallic deep purple sheen, creating a unique color unlike any other insect. The intricate wing veins reflect light in complex ways, making it shine like an amethyst over summer ponds.

隠形術

ツユムシ Katydid

世界各地の熱帯、それも湿潤で地衣類の多い地域には、地衣類を食べ、それに擬態したツユムシのなかまが生息している。本種はきわめて見事で、地衣類にとまると、まったく同化してしまうほどよく似ている。あきれるほど完成度の高い擬態である。こう見えて日本の普通のツユムシ類によく似た声で鳴く。

Some katydids feed on and mimic lichens in tropical regions rich in them. This particular species is remarkable for its perfect camouflage, blending so entirely with the lichens that it becomes nearly invisible. Its mimicry is impressively detailed. Interestingly, despite its appearance, it produces a call very similar to that of normal green katydids found in Japan.

直翅目／ウメノキゴケツユムシ *Gelotopoia bicolor* (Orthoptera: Tettigoniidae) (Cameroon). BL=18 mm.

［撮影メモ］地衣類に溶け込んで見える角度を探すのが大変でした。（法師人）

南の果ての山から

アフリカのクワガタ African stag beetle

アフリカ南部は独自の生物相をもつ地域で、多数の固有種が生息している。そのなかでも、テーブルマウンテンという山塊に生息するクワガタには奇妙なものが多い。マルガタクワガタのなかまは山頂ごとに異なる種が生息し、いずれも丸い体に奇妙な大顎や前脚の目立つ種ばかりである。本種はそのなかでも白眉で、まるでタラコのような大顎をもっている。その大顎には複雑な凹凸やねじれがあり、拡大してもまた奇妙である。

South Africa is home to many endemic species, including the unusual stag beetles that inhabit Table Mountain. Each peak hosts different species of Colophon beetles, all with round bodies and striking jaws or forelegs. This particular species stands out with reddish yellow strange mandibles. Upon closer inspection, these mandibles reveal intricate twists and ridges, making them even more fascinating.

鞘翅目／プリモスマルガタクワガタ *Colophon primosi* (Coleoptera: Lucanidae: Lucaninae) (Southern Africa). BL=31 mm.

砂漠のおもかげ

アフリカのゴミムシダマシ African darkling beetle

この種は石下やシロアリの巣に生息し、体から蠟状の物質を出し、周辺の基質を体につけることによって、地面と同化し、敵に気づかれないようにしているものと思われる。写真の標本はナミブ砂漠の近くの乾燥地で採集されたもので、拡大すると赤く美しい酸化鉄をまとった石英で覆われている。まるで砂漠の一部を背負っているかのようだ。ただの汚れにも思えてしまうが、生息地の環境を想像させるよう、標本がお土産をもっているかのようでもある。

This species lives under stones or in termite nests, secreting a waxy substance that helps it blend into its surroundings by attaching soil to its body, making it nearly invisible to predators. The specimen in the photo was collected in a dry area near the Namib Desert. Under magnification, it is covered in red quartz coated with iron oxide, resembling a piece of the desert itself. While it may appear as mere dirt, it gives a glimpse into the environment it inhabits.

鞘翅目／ゴミムシダマシの一種 *Eurychora* sp. (Coleoptera: Tenebrionidae: Pimeliinae) (Southern Africa). BL=11 mm.

[撮影メモ] 砂やごみを残しつつのクリーニングの加減が難しかったです。(吉田)

蜂の威を借る

南アメリカのツノゼミ

South American treehoppers

ハチに姿を似せる昆虫は少なくない。多くはハチの体の各部位に相当する部分をそのまま似せて擬態している。つまり、全身でハチに擬態している。カミキリムシなどでは、前翅の模様を腹部のしま模様に似せている例もあるが、概ね同じ部位を似せている。ところが、このハチマガイツノゼミは、前胸部分を巨大に発達させ、その部分がハチの体の大部分をつくり出している点で特異である。さらに鋭いトゲがたくさんあり、おそらく食べた捕食者は痛みを覚え、間違えてハチを食べたと確信することだろう。中央・南アメリカを代表する珍奇なツノゼミの一群である。

半翅目／ハチマガイツノゼミのなかま (Hemiptera: Membracidae: Heteronotinae) (South America). 1, *Heteronotus nigrogiganteus,* male; 2, *ditto,* female; 3, *H.* cf. *delineatus*; 4, *H. pompanoni*; 5, *H. horridus*; 6, *H. reticulatus*; 7, *H. brindleyae.*

Many insects mimic the appearance of bees and wasps, often copying corresponding body parts to create a full-body resemblance. For example, some longhorn beetles mimic the stripes on a wasps's abdomen. However, the bee-mimicking treehopper is unique in that it has an enlarged pronotum (the front part of its thorax), which makes up most of its wasp-like appearance. This group of strange treehoppers is a remarkable example of mimicry from Central and South America.

3
4
5
6
7

墓を背負う

日本のサシガメ 1　Japanese assassin bug 1

サシガメは他の昆虫をおそい、その体液を吸う肉食性のカメムシのなかまである。獲物は分類群によって異なり、決まったものを食べる種も多い。このハリサシガメはアリを食べる専門家である。そして面白いのは幼虫は死んだアリを背中に背負い、集め続けるのである。写真の幼虫は大量のトビイロシワアリの死体を背負っている。死体を好んで食べる生き物はいない。おそらくそれらを背負うことによって、敵に狙われにくくなるのだろう。ちょっと驚くような生態である。

Assassin bugs are predatory insects that feed on the body fluids of other insects, often specializing in specific prey. This particular species, the ant-eating assassin bug, specializes in hunting ants. Interestingly, the nymphs carry the bodies of dead ants on their backs, continuously collecting them. Since few animals prefer to eat dead insects, carrying these ant corpses may help the nymph avoid predators.

半翅目／ハリサシガメ *Acanthaspis cincticrus*
(Hemiptera: Reduviidae) (Japan). BL=5.4 mm.

［撮影メモ］背負っている死体同士の接着が脆いため、崩れないよう注意しながら撮影しました。（法師人）

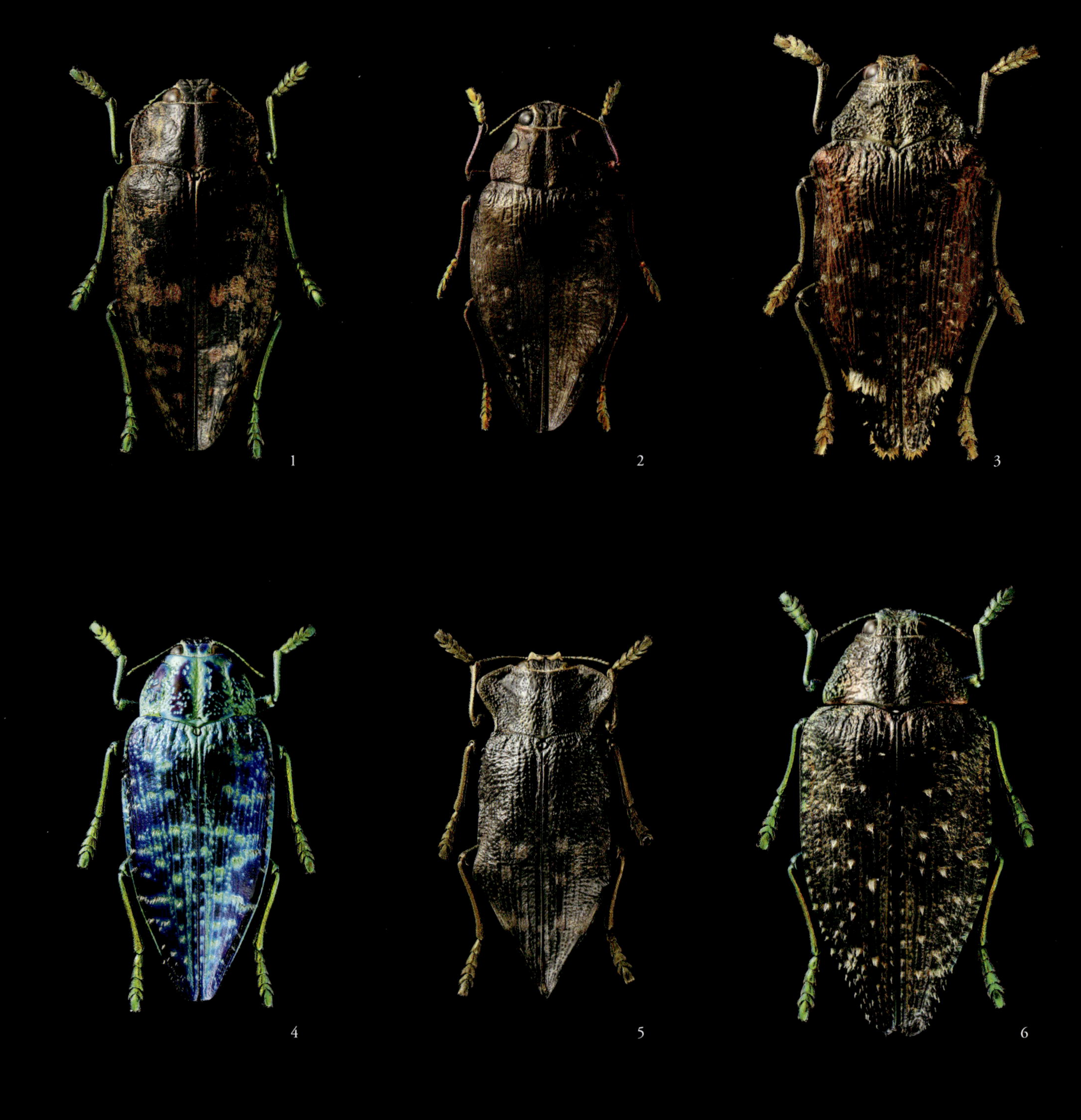

裏地のこだわり

マダガスカルのタマムシ

Maragasy jewel beetles

鞘翅目／カワリタマムシのなかま *Polybothris* spp. (Coleoptera: Buprestidae) (Madagascar). 1, *Polybothris infrasplendens*; 2, *P. muelleri*; 3, *P. scenica*; 4, *P. sumptuosa* var. "*superba*"; 5, *P. dilatata*; 6, *P. deyrollei*.

マダガスカル島固有のタマムシで、単純な紡錘形の種が多いタマムシとしてはさまざまな形のものがいることから、この名前がある。ここには、かなり珍しい種を厳選して図示した。一部の種は非常に派手だが、多くの種の背面はいぶし銀のような魅力がある。黒い背景だからこそ生きる色といえる。

This jewel beetle is endemic to Madagascar, a place known for its wide variety of shapes compared to the typically spindle-shaped jewel beetles. The species shown here are rare and carefully selected. While some are very colorful, many have a subdued charm with a burnished silver appearance. Their beauty is especially striking against the black background.

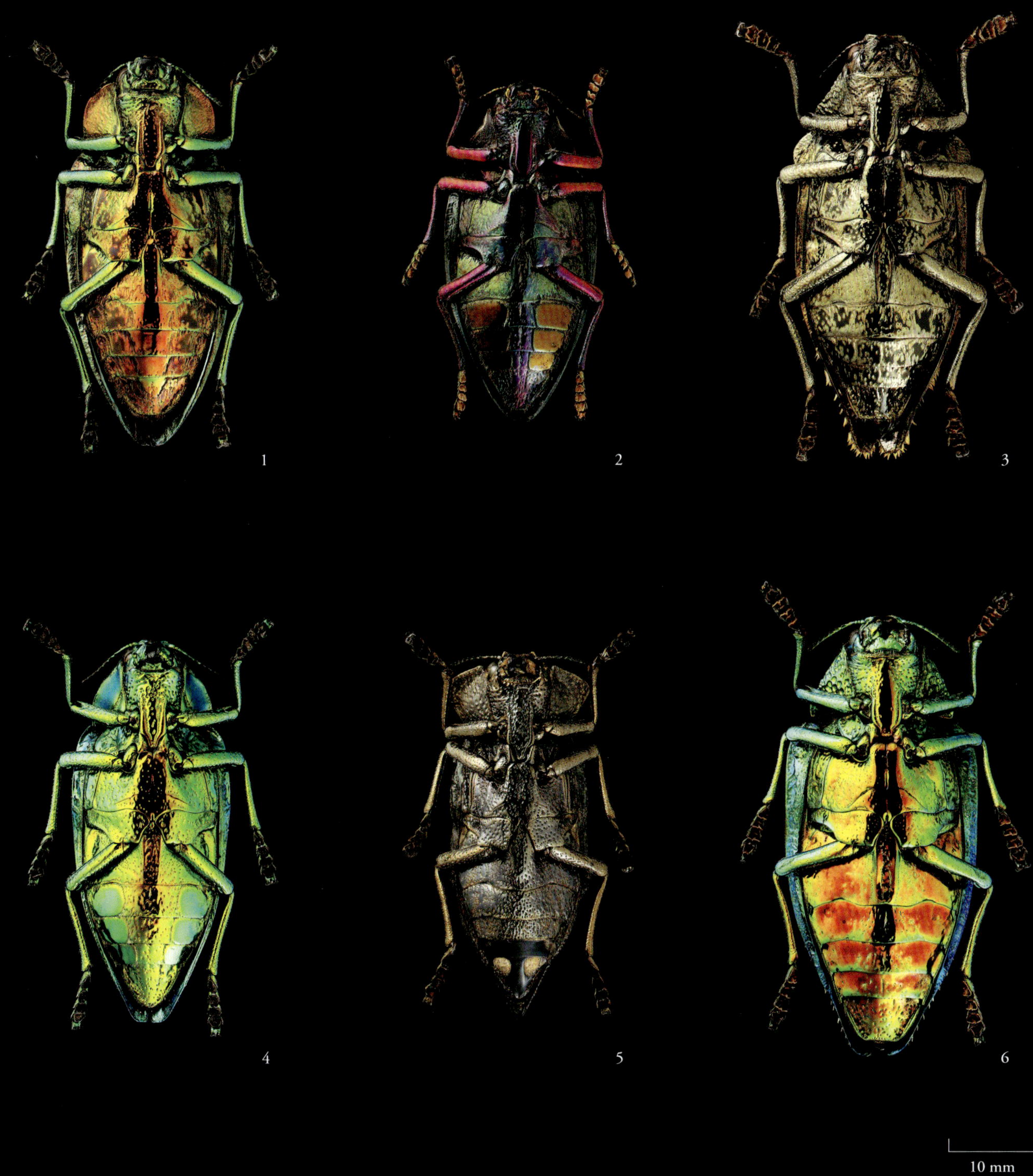

こちらは腹面である。背面が地味な種でも、腹面だけは派手な種がいる。これもまたカワリタマムシのなかまの面白さである。

These are the underside of the beetles. Some species with plain dorsal sides have surprisingly colorful undersides, adding to the intrigue of this unique group of jewel beetles.

キマイラ

カマキリモドキ Mantidfly

カマキリモドキはなんとも不思議な昆虫である。カマキリのような三角の頭と前脚（鎌）をもち、カマキリのように小さな昆虫を捕まえて食べる。その名の通りカマキリとは似て非なるもので、実際はウスバカゲロウなどに近い。面白いのは成虫の姿だけではない。なんと幼虫はクモの卵嚢に寄生するという奇妙な生活史をもつ。孵化したばかりの幼虫は脚が発達しており、まず通りがかったクモに抱きつく。次に、そのクモが運よく産卵すると、その卵嚢に乗り換えて、中身を食べて成長するのだ。きっと成功する確率は低いだろう。ウスバカゲロウのなかまとしては、非常にたくさんの卵を産む。

The mantidflies are fascinating insects allied to lacewings. It has a triangular head and raptorial front legs like a praying mantis, and it captures and eats small insects. What's even more interesting is its unique life cycle. The larvae parasitize spider egg sacs. Newly hatched larvae have developed legs and first cling to a passing spider. If the spider lays eggs, the larva transfers to the egg sac and feeds on the contents to grow.

脈翅目／オオイクビカマキリモドキ *Euclimacia badia* (Neuroptera: Mantispoidea: Mantispidae) (Japan). BL=26 mm.

ずんぐりとしてかわいらしいコガネムシである。この属の多くの種は開けた場所や乾燥地を好む。そのため、森林の多い日本では少なく、シロスジコガネを含む3種が海岸や河川敷に見られるのみである。拡大すると刺繍のような白い毛が繊細な模様をつくり出していることがわかる。そして触角は薄片が重なってできている。箱の写真はヨーロッパから日本にかけて生息する種を並べたものである。ここにはないが、アメリカにはシロスジコガネのなかまがとくに繁栄している。拡大写真は日本産のヒゲコガネ *Polyphylla laticollis* で、日本本土ではカブトムシに次ぐ大型のコガネムシである。

鞘翅目／ヒゲコガネ *Polyphylla* spp. (Coleoptera: Scarabaeidae: Melolonthinae). BL=ca. 25-45 mm.

These are stout and charming scarab beetles. Many species in this genus prefer open or dry areas, but forests cover Japan, and only three species are found mainly along coasts and riverbanks. Upon closer inspection, delicate white hairs form intricate, embroidery-like patterns on their bodies, and their antennae consist of overlapping plates. The pictured collection shows species from Europe to Japan. The magnified species is a Japanese species, *Polyphylla laticollis*.

刺繍

コガネムシ

Scarab beetles

つるはし

南アメリカのカミキリムシ

South American longhorn beetle

南アメリカを代表する奇怪な甲虫で、似たものがまったくおらず、何のなかまであるか昔から議論の的となってきた。実はカミキリムシの一種で、細長いひょうたん型の体に奇妙な形の脚が生えており、おかしな形の頭部をもっている。飛ぶことはできず、地中に彼らの生活圏があるといわれている。雨の後、地上を歩いているものがたまに見つかる程度で詳しい生態はまったくわかっていない。形の意味も謎であり、なにもかもが意味不明な虫なのである。

鞘翅目／ケラモドキカミキリ *Hypocephalus armatus* (Coleoptera: Vesperidae: Anoplodermatinae) (S. America). BL=52.5 mm.

This strange beetle from South America has no close relatives, leading to long-standing debates about its taxonomic position. It is a member of the longhorn beetles with a slender, gourd-shaped body, odd legs, and a peculiar head. Unable to fly, it is rarely seen above ground, usually only after rain, suggesting it lives mostly underground. Its behavior and the purpose of its unusual shape remain unknown, making this beetle a true mystery.

空っぽの天狗

大型のビワハゴロモ

Large lanternfly

その昔、ビワハゴロモの「鼻」のように見える長い頭部が光ると信じられたことがあり、そのため英名の「lanternfly」がある。しかし実際には光らない。とくに本種のように派手な色彩をもっていると、光らないかどうか期待してしまう昔の人の気持ちもわからないでもない。ときに眼より先の部分が折れている個体が見つかるが、驚くことに中身はほとんど空洞である。その役割はまったくもって不明であるが、少なくとも邪魔にならない程度に軽くできているのは間違いないようだ。本種はインドシナ半島の限られた地域にのみ生息する。

In the past, people believed that these bugs' long, nose-like heads could glow, which led to its English name, "lanternfly." However, it does not emit light. Sometimes, individuals are found with broken heads, revealing that the inside is mostly hollow. The exact purpose of this structure remains to be discovered.

半翅目／インドシナテングビワハゴロモ *Pyrops ducalis*
(Hemiptera: Fulgoridae) (SE Asia). Forewing length=34 mm.

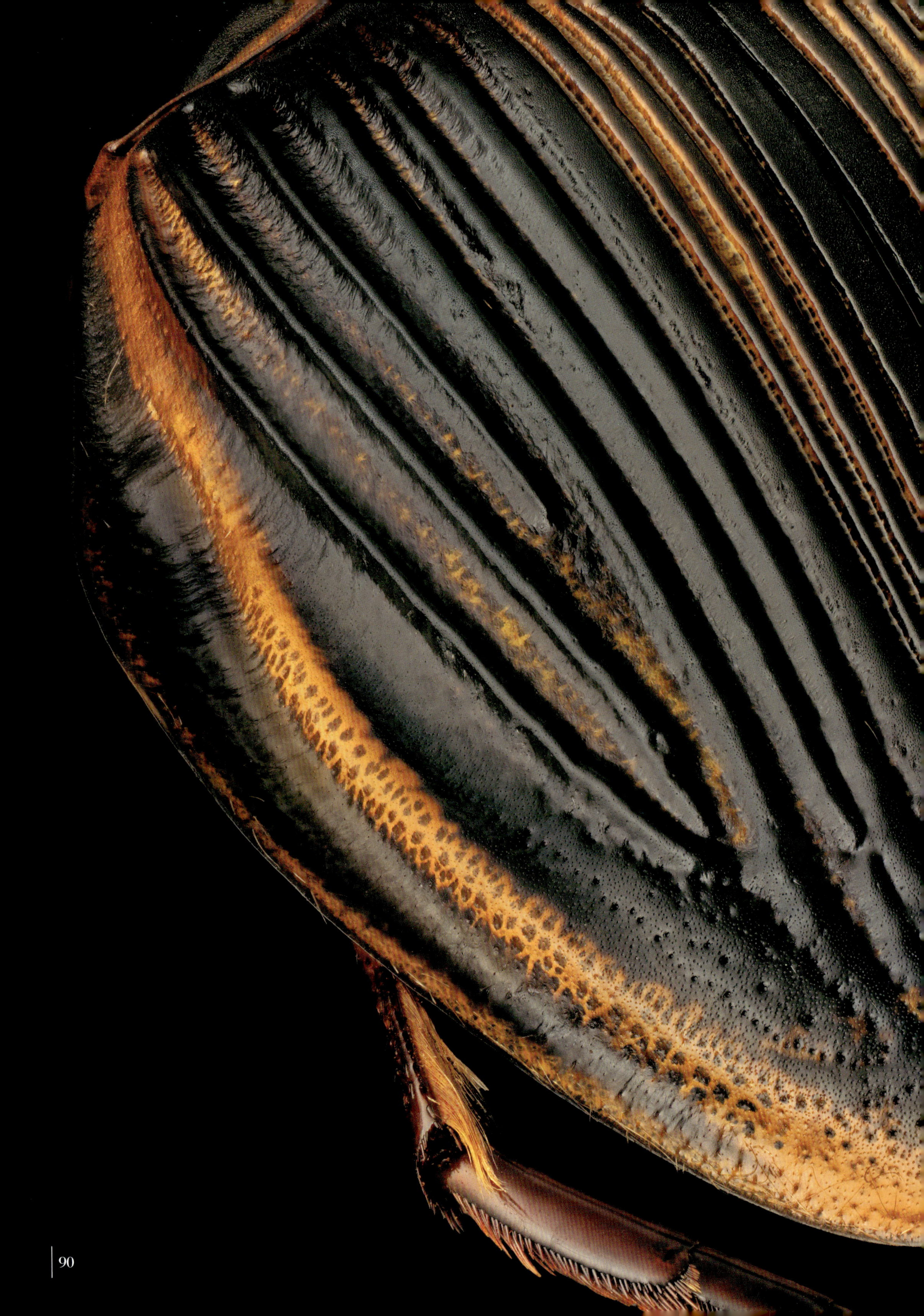

整流板

ゲンゴロウ Diving beetle

本種は世界最大級のゲンゴロウで、左右に張り出した前翅の形が異様な雰囲気を醸し出している。ヨーロッパの清浄で深い湖沼に生息し、そのような環境の開放的な水中で、安定した長距離の水泳を行う際に、このような姿が役立つものと思われる。大多数のゲンゴロウは浅い水域を好むので、その点でも本種は特異である。また、結氷期には穴釣りで釣れることもあるというから、きわめて低温に強いのであろう。環境開発や水質悪化により、多くの地域で絶滅している。冷涼な水中を好む本種にとって、近年の温暖化も重大な問題だろう。この標本は絶滅したドイツで1800年代終わりに採集されたものである。

鞘翅目／オウサマゲンゴロウモドキ *Dytiscus latissimus* (Coleoptera: Dytiscidae; Dytiscinae) (Europe). BL=44 mm.

This species is one of the world's largest diving beetles, with an unusual elytral shape that extends outward, creating a distinct appearance. It lives in Europe's deep, clean lakes, where this body shape likely aids in steady, long-distance swimming in open water. Unlike most diving beetles, which prefer shallow areas, this species is unique. However, environmental development and water quality decline have led to its extinction in many regions. This specimen was collected in Germany in the late 1800s, before its extinction there.

蓬に化けて

ヤガ Noctuid moth

秋にヨモギの花を見てまわるとこのイモムシが見つかる。ヨモギの花序の輪郭とその色合いを忠実に再現し、風景に溶け込んでしまう。見事な擬態である。晩秋に蛹となり、そのまま冬を越し、さらに春と夏も蛹のまま過ごし、秋口に成虫となって産卵する。成虫はその名の通り灰色で木目がある。地味ながら味わい深い模様の蛾である。

In autumn, you can find this caterpillar on the flowers of mugwort. It blends perfectly with the shape and color of the flower clusters, making it an excellent example of mimicry. The caterpillar pupates in late autumn, then overwinters and remains a pupa through spring and summer, emerging as an adult moth in early autumn. The adult moth is gray with woodgrain-like patterns, creating a subtle yet appealing design.

ハイイロセダカモクメ *Cucullia maculosa* (Lepidoptera: Noctuidae) BL=12 mm (larva); forewing length=19 mm(adult).

視点を変えて

カナブン Drone beetle

アフリカはハナムグリ亜科のコガネムシ（カナブンとハナムグリ）の宝庫である。どうしてかというと、同じ枯れた木を幼虫が食べるカブトムシやクワガタが少ないからだと考えられている。アフリカの大部分の地域は乾燥しているが、そのような場所にもカナブンは強いというのもあるだろう。発達した角をもっていて、樹液に集まってカブトムシのように戦うものも少なくない。本種もそのような種で、立派な角をもっている。多くのカナブンの表面は、コレステリック液晶という構造によって、光の当たる角度と見る角度の関係でさまざまな色に変化する。地域によって多くの変異があるが、この個体の産地のものはとりわけ美しい。

Africa is a haven for flower beetles (Cetoniinae), a subfamily of scarab beetles. Some species, like this one, have large horns and engage in battles over tree sap. The surface of many flower beetles features a cholesteric liquid crystal structure, causing their colors to change depending on the angle of light and view. This particular individual is from a region known for its especially beautiful variations.

鞘翅目／ミツノカナブン *Neptunides polychrous* (Coleoptera: Scarabaeidae) (Southern Africa). BL=30 mm.

［撮影メモ］生きているときは俊敏で、撮影には麻酔が必要ですが、それが効きすぎると脚を伸ばして倒れてしまう。程よい状態で撮影するのに苦労しました。（丸山）

幅を利かせる

シュモクバエ Stalk-eyed fly

名前は仏具の鐘をたたくハンマーである撞木に由来する。たしかに背中から見るとそのような形で、なんとも不思議である。この形の理由としてあげられるのは、オス同士がメスをめぐって争う際、頭を突き合わせ、眼の間隔（頭の幅）の広いほうが勝つという説がある。生きるのに不便そうな形だが、それこそ強いオスの証であり、そのため眼の間隔が広いほうがメスに好まれるという説もある。

This odd form may serve a purpose: when males compete for females, they clash heads, and the one with the wider head (or eye span) often wins. While this shape may seem impractical, it is thought to signal strength, making males with wider heads more attractive to females.

双翅目／ニシアフリカシュモクバエ *Diopsis apicalis* (Diptera: Diopsidae) (C. Africa). BL=6.5 mm.

二つ折りの扇子

ハサミムシ Earwig

コブハサミムシは山地性の種で、春先に渓流沿いの石の下で、メスが卵を守っている様子を観察することができる。そして孵化した幼虫は母親の体を食べて巣立つ前の食料とするという興味深い生態をもつ。ハサミムシの面白さはその形にもある。腹部末端の鋏の造形も見事だが、後翅の折りたたみ構造が素晴らしい。ちょうど扇子を途中で折り曲げたようなたたみ方で、実に機能的にできている。この仕組みをなにかに利用できないかという研究も行われている。左の写真は、翅を開いて拡大したものである。

革翅目／コブハサミムシ *Anechura harmandi* (Dermaptera: Anisolabididae) (Japan). BL=18 mm.

This earwig species lives in mountainous areas, and in early spring, you can observe females guarding their eggs under stones near streams. Interestingly, the newly hatched larvae feed on their mother's remains before leaving the nest. Earwigs' unique shape is also fascinating. While the pincers at the end of the abdomen are impressive, its wing-folding mechanism is remarkable, resembling a folded fan. This functional design has inspired research on folding satellite antennas and solar panels.

1

2

3

皇帝の風格

南アメリカのゾウムシ South American weevils

世界各地の熱帯には必ず美しいゾウムシが生息している。いずれも美しいと思える派手な色彩は、硬さを天敵に誇示するための警告であり、それが各地のゾウムシで独自に進化を遂げている。南アメリカを代表するゾウムシはこれらコウテイゾウムシのなかまである。一部の種は非常に大型で、金緑色の鱗毛をちりばめて出来上がった見事な色彩を示す。また脚ががっちりとして毛深く、脚の先端（跗節）が大きいのも特徴的で、全体に貫録がある。1種は皇帝という意味の種名をもつが、その姿に昔の人も皇帝の印象を重ねたのであろう。

鞘翅目／コウテイゾウムシ (Coleoptera: Curculionidae: Entiminae). 1, *Entimus imperialis*; 2, *E. granulatus*; 3, *E. splendidus* (S. America). BL=28 mm (*E. imperialis*).

Tropical regions worldwide are home to beautiful weevils. These weevils often display bright colors that serve as a warning to predators and showcase their toughness. These striking colors have evolved independently in different species. In South America, the *Entimus* weevils stand out. Some species are large, with brilliant colors created by golden-green scales. Their sturdy, hairy legs and large tarsi add to their imposing appearance.

ミクロのブルドーザー

キクイムシ Bark beetle

キクイムシは筒状の面白い形をした甲虫である。ゾウムシ科に含まれるが、ゾウムシに特徴的な長い口吻をもたない。どうしてそのような形をしているかというと、枯れた木材に坑道を掘って生活するからである。木材を食べるほか、一部の種では坑道に菌を植え付け、それを増やして餌としている。オスとメスで形に差があるものが少なくなく、おそらく坑道（なわばり）やメスをめぐって熾烈なオス同士の争いがあるのだろう、オスに小さな角があるものも多い。本種は前胸部全体がスコップのような形をしており、非常に変わった形をしている。おそらく狭い坑道のなかでこの部分をぶつけあってオス同士が戦うのであろう。

Bark beetles are tube-shaped beetles that belong to the weevil family, but unlike weevils, they don't have long snouts. Their shape helps them dig tunnels in dead wood where they live. Some species of males often have small horn(s), likely used in fights over tunnels (territory) or mates. This particular species has a shovel-like pronotum, a unique feature probably used by males to clash with each other in narrow tunnels.

鞘翅目／キクイムシの一種 *Eccoptopterus* cf. *spinosus* sp., male (Coleoptera: Curculionidae:Scolytinae) (SE Asia). BL=4.1 mm.

歴史の証明

ムネアカセンチコガネ Dor beetles

ムネアカセンチコガネのなかまは日本を含む世界中に分布し、成虫幼虫ともに、一部の種で地下の菌類を食べていることがわかっている。乾燥地を好むため、アフリカにはとりわけ多くの種が繁栄している。この標本の下には多くのラベルが刺さっている。1枚目のラベルは、この個体がベルギー人のF. G. Overlaetによって1933年に採集されたことを示す。2枚目によりフランス人のA. Boucomontがこの個体を*Bolbaffer princeps*と同定したことがわかる。ドイツ人のR. Petrovitzは*B. gigas*であると異論を述べたようだが、ロシア人のG. V. Nikolajevは*B. princeps*で間違いないとしている。いずれも高名な昆虫学者たちである．彼らの声が、茶色く変色したラベルから聞こえてくる。右頁はこれと同じような歴史的標本を集めたものである。

鞘翅目／サバンナセンチコガネ *Bolbaffer princeps* (Coleoptera: Bolboceratidae) (Southern Africa). BL=22 mm.

Dor beetles are distributed worldwide, with some species feeding on underground fungi as adults and larvae. These beetles thrive in dry areas, particularly in Africa. This specimen has several labels. The first label shows it was collected by Belgian entomologist F.G. Overlaet in 1933. The second indicates that French entomologist A. Boucomont identified it as *Bolbaffer princeps*, while German R. Petrovitz argued it was *B. gigas*. However, Russian entomologist G.V. Nikolajev confirmed it as *B. princeps*. These renowned entomologists' voices seem to echo from the faded labels.

名もなき華麗

日本のサシガメ 2 Japanese assassin bug 2

カメムシは甲虫やハチなどにくらべ、その多様性の割には、世界的に研究者が少ない。日本産の比較的大型の種でも、分類学的に問題の残っている種や、学名に問題のある種がまだ多く残されている。本種は日本産の大型種で、しかもきわめて大型美麗な種として、名前の明らかになっていない種の筆頭であろう。琉球列島で枯れ木で発見されているが、驚きの完成度を誇る華麗ともいえる地衣類擬態で、見つけるのは難しい。拡大すると大理石のような見事な彫刻が浮かび上がる。

半翅目／ハラビロトゲサシガメ *Neocentrocnemis* sp. (Hemiptera: Reduviidae) (Japan). BL=22 mm.

Despite their diversity, true bugs are less studied than beetles or wasps, with few researchers worldwide. Even among Japan's larger species, many still need help with taxonomic issues or incorrect scientific names. This particular species, one of Japan's largest and most beautiful, is still unnamed. Found on dead wood in the Ryukyus, its excellent lichen mimicry makes it hard to spot. When viewed up close, its intricate, marble-like patterns are awe-inspiring.

畏怖すべし

スズメバチ Hornet

世界でもっとも恐ろしいハチであることは論を俟たない。非常に大型かつ攻撃的で、毒も強い。実際、毎年このハチに刺されて亡くなっている人がおり、恐ろしさの質は異なるが、クマより被害が大きいとされる。他の昆虫を捕まえて食べ物にするほか、樹液などにも集まり、そこを独占しようとする。餌場を縄張りにする数少ないスズメバチでもある。その恐ろしい実態もあいまって、姿も見事なハチである。がっちりとした体躯に大きな頭をもつ。毒針（右頁）は長大で、刺されると奥まで毒が注入されることを思うと、見るだけで冷や汗が出る。

膜翅目／オオスズメバチ *Vespa mandarinia* (Hymenoptera: Vespidae) (Japan). BL=33 mm.

This species is undoubtedly one of the world's deadliest wasps. It is huge, aggressive, and highly venomous. In fact, people die from its stings every year, and it causes more harm than bears. In addition to hunting other insects, it gathers around tree sap and fiercely defends its food source. Its imposing body and large head make it both intimidating and striking in appearance. Its long stinger injects venom deeply, making it a truly terrifying insect.

ジュリエットの香り

ホノオムシ Glowworm beetle

ホノオムシ科は新世界に生息するホタル科に近い甲虫の一群である。写真はオスの成虫で、メスは幼虫型（イモムシ型：幼形成熟）である。幼虫とメスの成虫はすべての体節に発光器官をもち、夜行列車のように光る。幼虫はヤスデを食べることがわかっている。夜行性で、オスは匂いと発光の光を頼りにメスを探す。オスの触角は非常に細かく枝分かれしており、表面には匂いを感じる感覚器にびっしりと覆われている。これによってメスの微量なフェロモンを捉える。

鞘翅目／ホノオムシの一種 *Phengodes* sp. (Coleoptera: Phengodidae) (N. America). BL=14 mm.

The Phengodidae are a group of beetles related to fireflies found in the New World. The photo shows an adult male while the female remains larval in form (neotenic). Both larvae and adult females have light-emitting organs on all body segments, glowing like a nighttime train. The larvae are known to feed on millipedes. Males are nocturnal and use scent and light to find females. Their antennae are finely branched and densely covered with sensory organs to detect the faint pheromones released by females.

風雅

日本のタテハチョウ Japanese Nymphalidae

日本の昆虫でもっとも風情のある和名のひとつだといわれている。墨を水面にたらし、そこに紙や布をあてることによって、波のような模様をつける「墨流し」に由来する。たしかにそのような見事な模様である。昔の人がつけたからこそ、きっとこのような語彙が選ばれたのだろう。今の時代であれば、タテハチョウのなかまであることから、「何とかタテハ」と風情のない名前になっていたと思われる。模様の墨にあたる部分が上品に青く輝き、その名前にふさわしい風情のあるチョウである。

鱗翅目／スミナガシ *Dichorragia nesimachus* (Lepidoptera: Nymphalidae) (Japan). Forewing length=32 mm.

This species has one of the most elegant Japanese names among insects. It is inspired by the traditional dyeing technique "Suminagashi," where ink is floated on water to create wave-like patterns on fabric. The butterfly's exquisite markings indeed resemble this art form. The dark "ink" parts of its pattern shimmer with a refined blue, making this butterfly genuinely deserving of its graceful name.

丸まる宝石

セイボウ Cuckoo-wasps

セイボウはハチのなかでもとりわけ美麗種の多い一群で、ハチとしては珍しく、メスも毒針をもたない。その代わり体は非常に硬く、おそらく小鳥などの捕食者に対して、この色彩は警告色として働くのであろう。そしてここに図示した通り、丸まって体を防御するのも面白い。多くの種は他のハチの巣に寄生する。そのため、英語では「カッコウのハチ」と称されるほか、美しさから「エメラルドのハチ」と呼ばれることもある。

Cuckoo-wasps are among the most beautiful wasps. Unlike most wasps, females lack a stinger. Instead, their bodies are very rigid, and their bright colors likely serve as a warning to predators like birds. As shown in the photo, they can also curl up defensively. Many species of cuckoo-wasps are parasitic on other wasps.

膜翅目／セイボウのなかま (Hymenoptera: Chrysidae: Chrysinae) (Japan). 1, **セイボウの一種** *Chrysis* sp. (BL=9.3 mm); **2, ムツバセイボウ** *C. fasciata* (BL=11.5 mm); **3, オオセイボウ** *Stilbum cyanurum* (BL=17 mm).

1

2

3

消えゆく漆黒

ナナフシ Stick insect

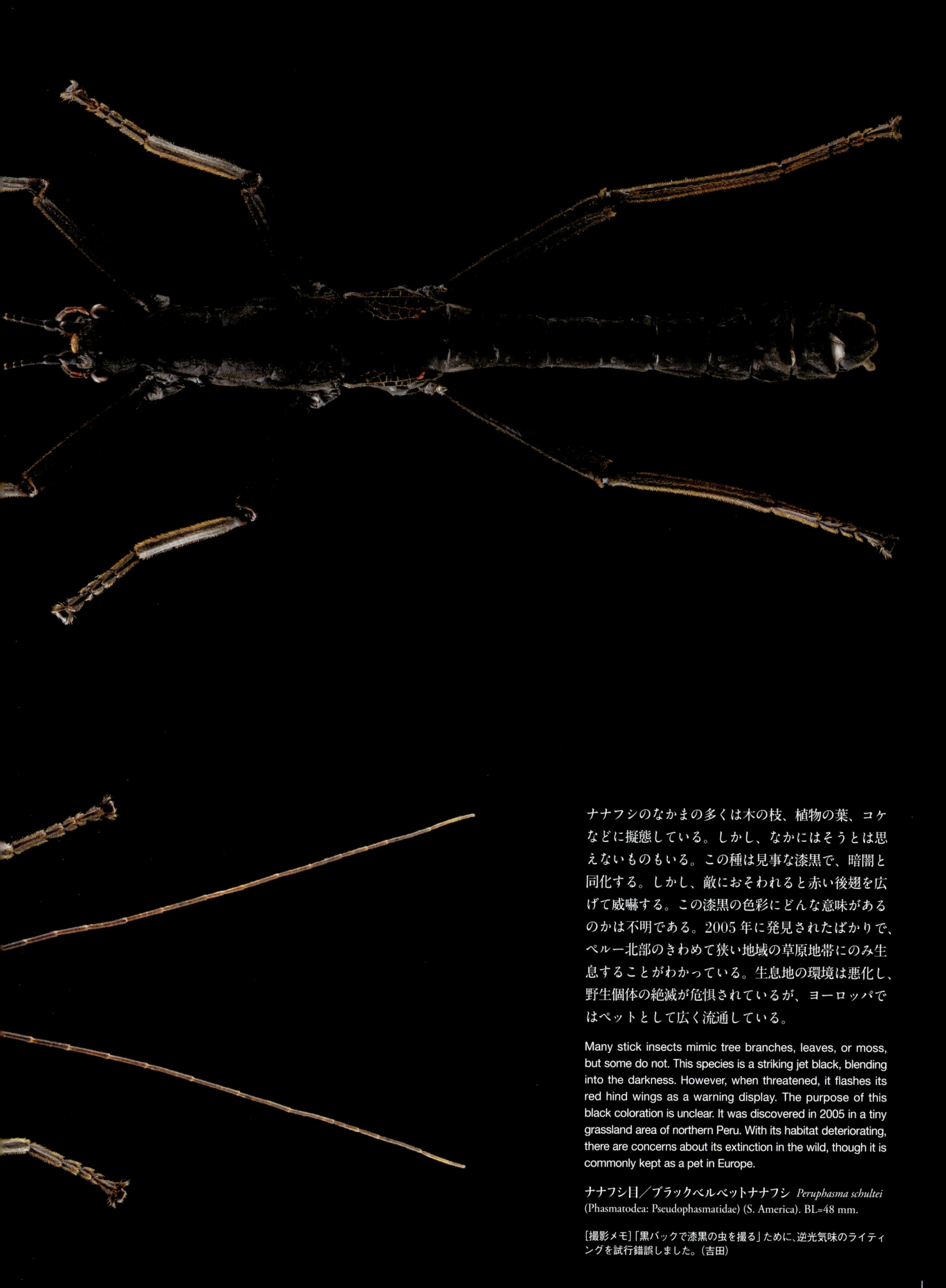

ナナフシのなかまの多くは木の枝、植物の葉、コケなどに擬態している。しかし、なかにはそうとは思えないものもいる。この種は見事な漆黒で、暗闇と同化する。しかし、敵におそわれると赤い後翅を広げて威嚇する。この漆黒の色彩にどんな意味があるのかは不明である。2005年に発見されたばかりで、ペルー北部のきわめて狭い地域の草原地帯にのみ生息することがわかっている。生息地の環境は悪化し、野生個体の絶滅が危惧されているが、ヨーロッパではペットとして広く流通している。

Many stick insects mimic tree branches, leaves, or moss, but some do not. This species is a striking jet black, blending into the darkness. However, when threatened, it flashes its red hind wings as a warning display. The purpose of this black coloration is unclear. It was discovered in 2005 in a tiny grassland area of northern Peru. With its habitat deteriorating, there are concerns about its extinction in the wild, though it is commonly kept as a pet in Europe.

ナナフシ目／ブラックベルベットナナフシ *Peruphasma schultei*
(Phasmatodea: Pseudophasmatidae) (S. America). BL=48 mm.

[撮影メモ]「黒バックで漆黒の虫を撮る」ために、逆光気味のライティングを試行錯誤しました。(吉田)

色とりどりの歴史

カイコ Silkworms

カイコは約6000年前に中国で絹をとるためにクワコから家畜化された昆虫である。世界中で飼育され、養蚕業は明治時代の日本で国の経済を支える一大産業ともなった。長年にわたり各地で品種改良が行われ、系統ごとに幼虫の性質や繭の形態などが異なる。九州大学は世界から収集されたもっとも多くのカイコの系統を維持しており、ワクチン開発や遺伝子工学などの先端研究にも利用されている。写真は色とりどりの繭で、それぞれに大きさや形、風合いが異なるのがわかるだろう。右下は成虫で、家畜化が進められたカイコと原種のクワコのメスを並べた。

In the Meiji period, silk farming became a primary industry supporting Japan's economy. Over time, selective breeding has produced many strains, each with unique characteristics in larval traits and cocoon shapes. Kyushu University maintains the most extensive collection of silkworm strains collected worldwide, which are used in advanced research like vaccine development and genetic engineering. The photo shows a variety of colorful cocoons, each differing in size, shape, and texture.

鱗翅目／カイコ Silkworm, *Bombyx mori* (Lepidoptera: Bombycidae) (breeding). **1, 濃山吹色繭** (S01: イタリア在来); 2, **淡黄繭** (c15: 自然突然変異); 3, **ぼか繭** (u90×p50：系統維持用雑種で、このように表面が綿状の繭をぼか繭という); 4, **緑繭** (c53: 青白という愛称の育成品種); 5-8, **紅繭 / 肉色繭** (c42-44: イタリア在来); **9, 多蚕繭 (b20:** 琉球多蚕繭ともよばれ、複数の幼虫が協働で繭をつくる **); 10, 鐘音** (p42: カネボウが改良した品種); 11, **浮ちぢら繭** (t11: 楕円形で、単一遺伝子 *Fl* で制御されるぼか繭を浮ちぢらと呼ぶ); 12, **白繭** (l82: 黄体色抑圧突然変異といい、レモン色幼虫のレモン色遺伝子が抑制され、白色の幼虫・繭となる); 13, **蛍光繭**（**櫻 3b：東大 (東京農大)・農研機構が開発した遺伝子組換えカイコで、紫外線を当てると繭が蛍光を発する**); 14, **成虫** (メス); 15, **原種クワコ** (メス).

10
11
12
13
9
14
15
10 mm

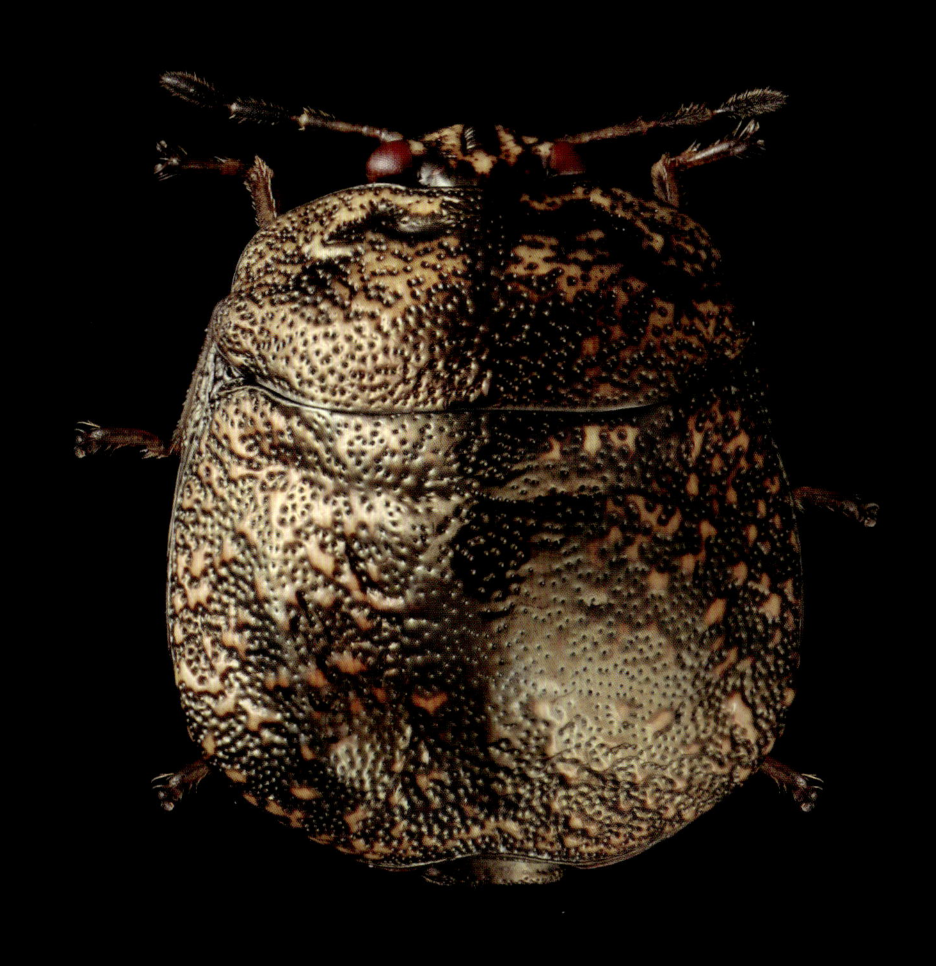

かくれ身、かわり身

日本のマルカメムシ Japanese globular stink bug

マルカメムシ科は日本に10数種が知られており、いずれも成虫は似たり寄ったりの姿だが、幼虫の形は面白いものが多い。サメハダマルカメムシは琉球列島に生息する未記載種（新種）で、幼虫（左頁）の異様さは際立っている。きわめて扁平で、レースのような縁取りがある。生木の樹皮にぴったりと張り付いており、それが地衣類のついた白っぽい樹皮上となると見つけるのが非常に難しい。成虫になるとまったく異なる形になるうえ、大きさもかなり小さくなってしまうのが不思議である。成虫（右頁）はその名の通りゴツゴツとした鮫肌で、銅色の体色が古びた硬貨のようで味わい深い。

The globular stink bug family Plataspidae includes around a dozen species known in Japan. The nymphs of this species are incredibly flat, with lace-like edges. They cling tightly to tree bark, making them very difficult to spot, especially on pale bark covered in lichens. Interestingly, the adult looks entirely different, becoming much smaller in size.

半翅目／サメハダマルカメムシ *Phyllomegacopta* sp. (Hemiptera: Plataspidae) (Japan). BL=8.5 mm (nymph); 6.4 mm (adult).

［撮影メモ］幼虫を撮影しようと思ったら、小さな成虫が羽化していてびっくり。あわてて追加の幼虫を探しました。幼虫には赤いダニがついています。（丸山）

鏡の真実

シャチホコガ Puss moth

スズメガに似ているがスズメガではなく、銀色の紋があるので、この名前がある。灯火にこの蛾がやってくるとき、その光を反射して美しく輝く。腹部の末端にある毛の束も見事で、美しいというより、奇妙なこの蛾を印象づける。銀色の紋を拡大すると、透明の鱗粉が重なってできていることがわかる。この構造で引き起こされる乱反射が輝きの正体である。この紋には蛾の姿を不明瞭なものとし、天敵である小鳥などに対し、生き物であることを認識させにくくする効果があるのだろう。

When this moth is attracted to light, it reflects the glow and shines beautifully. Its bundle of hairs at the tip of its abdomen adds to its unusual appearance. Upon closer inspection, the silvery marks are made of overlapping transparent scales, creating the shimmering effect through light scattering. These markings likely help obscure the moth's outline, making it harder for predators like birds to recognize it as a living creature.

鱗翅目／ギンモンスズメモドキ *Tarsolepis japonica* (Lepidoptera: Noctuoidea: Notodontidae) (Japan). Forewing length=68.5 mm.

曜変天目

日本のタマムシ Japanese jewel beetle

属名の *Nipponobuprestis* は「日本のタマムシ」を意味する。日本には２種がいるが、どちらも非常に美しく、その点をふくめて日本を代表するタマムシといえるだろう。本種は金色と黒を繊細なグラデーションで繋げ織りなす、曜変天目茶碗のようなまだら模様をもつ。エノキの古木に稀に見つかり、美しさと珍しさを兼ね備えている点もまた魅力である。

鞘翅目／クロマダラタマムシ *Nipponobuprestis querceti* (Coleoptera: Buprestidae) (Japan). BL=22 mm.

The genus name *Nipponobuprestis* means "Japanese jewel beetle." Japan is home to two species, both exceptionally beautiful and representative of Japan's jewel beetles. This species features a delicate gradient of gold and black, reminiscent of the mottled patterns of the famous Yohen Tenmoku tea bowls. These rare beetles are found on old hackberry trees, making them prized for their beauty and rarity.

浮遊の翅

アザミウマ Thrip

アザミウマは小型種の多い昆虫の一群で、多くは植物の汁を吸い、ときに農作物の大害虫となる。2ミリメートル程度のものが多く、肉眼ではなかなか気づかないが、実は植物のあるところであれば、どこにでもいる昆虫でもある。微小種が多いアザミウマにあって、大型種が多いのが本種を含むオオアザミウマのなかまで、ときに7ミリメートルを超えることもある。本種はオスの腹部の途中に特徴的なトゲがあり、幼虫が鮮烈な赤色なのも面白い。枯れ葉におり、菌類の胞子を食べる。アザミウマの多くは翅が繊細な羽毛状で、本種も同様の翅をもつ。ある一定以下のサイズになると、空気中を漂うように飛行するほうが効率的であるといわれ、そのためこうした羽毛状の翅をもつ微小昆虫がさまざまな分類群に見られる。

Thrips are a group of small insects, many of which feed on plant sap and can become severe pests to crops. Most species are around 2 millimeters in size and often go unnoticed, but they are found wherever there are plants. This species is large, with males featuring distinctive spines on their abdomens. The larvae are strikingly red, which is an exciting trait. Like most thrips, they have delicate, feather-like wings to waft in air.

総翅目／オオトゲクダアザミウマ *Bactrothrips honoris* (Thisanoptera: Phlaeothripidae) (Japan). BL=6.5 mm (adult); 4.2 mm (nymph).

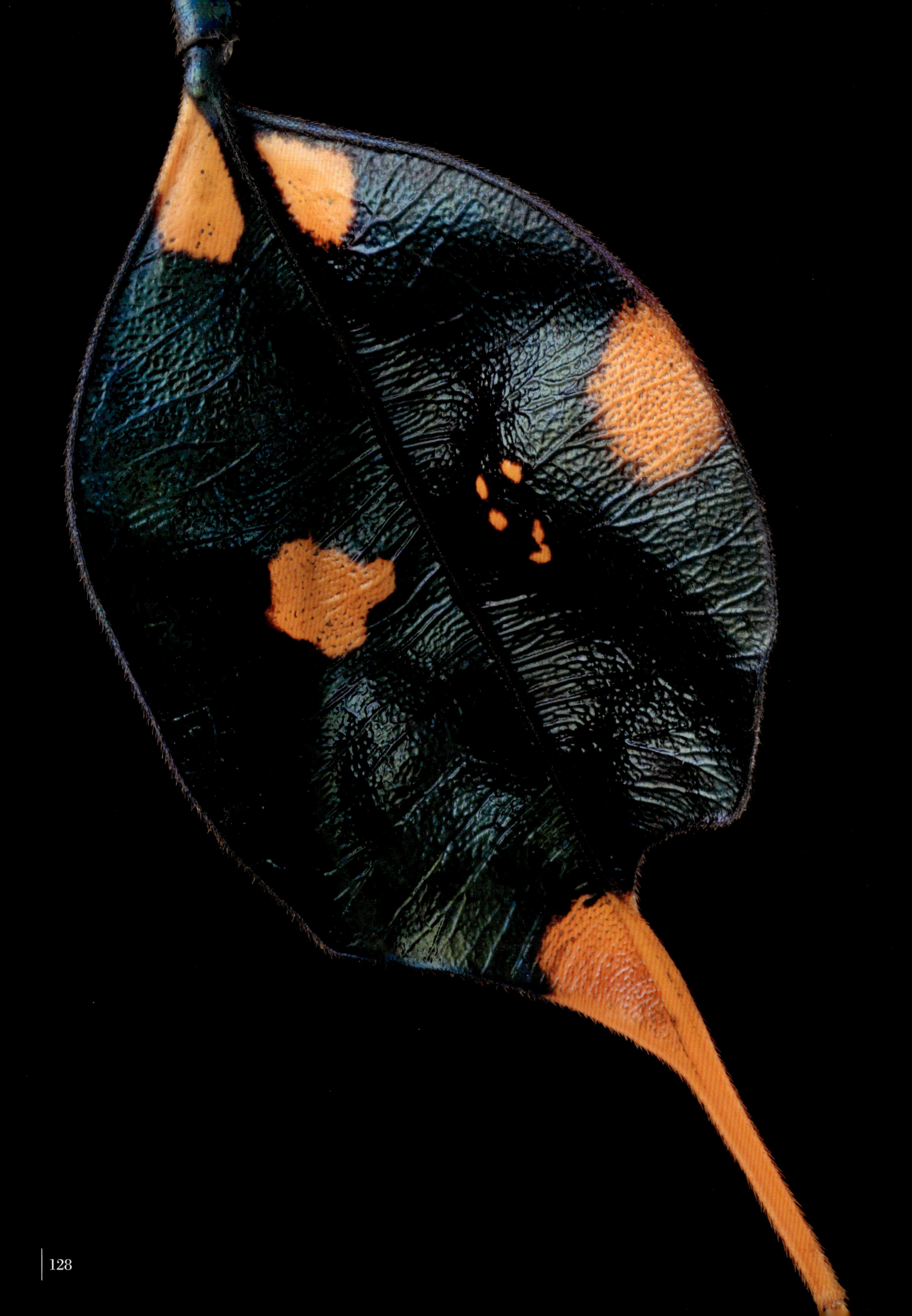

道化師

ヘリカメムシ Squash bug

一見してとても変わったカメムシである。後脛節の先に「軍配」をもち、それが鮮やかな色彩を呈している。さらにその部分は、平面的に見えるが実は立体的で、陽光下できらきらと輝く。果たしてこの形と色にどんな意味があるのだろうか。毒があることを示すのであれば別の部分を目立たせればよいし、これがあることによって小鳥などが食べにくくなるとも思えない。ひとつわかることは、この部分はきわめてもろく、少しつついただけでも破れてしまう。なにか生存に役立つのかもしれないが、実際の機能は不明なまま。美しくも悩ましい形態である。

This bug has brightly colored "fans" at the end of its hind legs, which appear flat but are three-dimensional and shimmer in the sunlight. The purpose of this shape and color needs to be clarified. While it could serve as a warning for predators, other body parts might be more suitable. Though its function remains unknown, this beautiful structure is intriguing.

半翅目／ニシキグンバイヘリカメムシ *Diactor bilineatus* (Hemiptera: Coreidae) (S. America). BL=20.5 mm.

蠱毒

ツチハンミョウ

Blister beetle

ツチハンミョウのなかまは寄生性で、幼虫はハチの巣に集められた餌や他の昆虫の卵塊を食べて成長する。本種のようにハチの巣に寄生する場合、生まれたばかりの幼虫は長い脚に発達した爪をもち、花などで待ち伏せし、運良く寄主のハチが来れば、それにしがみつき、巣に運ばれる。そして成長の過程で、コガネムシ幼虫状の姿となり、まったく動かない蛹のような幼虫期(擬蛹)を経て、再びコガネムシ幼虫状の姿で成長し、蛹となる。このような複雑に姿を変える成長過程を過変態と呼ぶ。また、最初にハチに運ばれる確率は非常に低く、そのためメスは数千個から１万個の卵を産む。ふくよかなこの姿は大量の卵を抱えているためである。このなかまは一様に強い毒をもち、この美しくも毒々しい姿はそのことを示す警告色である。

鞘翅目／ニセツチハンミョウの一種 *Pseudomeloe* sp. (Coleoptera: Meloidae: Meloinae) (S. America). BL=30 mm.

Blister beetles are parasitic; their larvae feed on other insects' egg mass or food stores of bee nests. When parasitizing bee nests, like this species does, the newly hatched larvae have long, specialized claws to cling to bees and hitch a ride back to their nests. Through growth, they transform into a larval form resembling a scarab larva, go through an inactive stage (coarctate larva), and then develop further before pupating. This complex development is called hypermetamorphosis. The likelihood of reaching a bee's nest is very low, so females lay thousands of eggs—up to 10,000. Their plump appearance is due to carrying many eggs.

緑の韋駄天

ゴキブリ Cockroach

本種はおそらく世界でもっとも美しいゴキブリのひとつだろう。そうでありながら、形は典型的なゴキブリのままというのも面白い。緑色の金属光沢を放ち、黄色い斑紋をもっている。学名の種小名は「タマムシのような」という意味で、まさにそのような色彩である。晴れた日に林縁や草むらの植物の上に見られ、きわめて敏速に飛び回り、採集するのは難しい。捕まえても網からいつの間にか消えてしまう素早さで、こんなゴキブリは初めてだった。色彩をふくめて謎が多い。

ゴキブリ目／タマムシモドキゴキブリ *Oxyhaloa buprestoides* (Blattodea: Blaberidae: Oxyhaloinae) (W. Africa). Left: male (BL=16 mm); right: female (BL=20 mm).

This species is likely one of the most beautiful cockroaches in the world, yet its shape remains that of a typical cockroach. It has a metallic green sheen with yellow spots. Its species name means "like a jewel beetle," perfectly describing its vibrant colors. On sunny days, it can be found on plants at the edge of forests or in grasslands, flying quickly and making it difficult to catch. Its coloration and behavior remain mysterious.

［撮影メモ］麻酔して撮影しました。（丸山）

閉じ込められた時間

ツノゼミの化石と現生種

Fossil treehopper and extant species

ツノゼミは南アメリカ周辺で生まれ、多様化した一群とされている。少なくとも1億年前の化石群からは見つかっておらず、それ以降に生じた比較的新しい分類群であることは間違いない。写真は中央アメリカのドミニカ琥珀に入っているツノゼミの化石である。ドミニカ琥珀は研究者によって考えが違うが、2000万年〜4000万年前のものとされている。この化石の面白い点は、明らかな近縁種（標本写真）が現在のアジアに見られることである。それに対して、中央アメリカや南アメリカには近い種はおらず、絶滅してしまったと考えられる。分類群の分散や盛衰の歴史を語る重要な資料である。

Treehoppers are thought to have originated in South America, with no fossils older than 100 million years, indicating they are relatively recent. The photo shows a treehopper fossil in Dominican amber, estimated to be 20 to 40 million years old. Surprisingly, a closely related species is found in Asia today, while none remain in Central or South America, suggesting extinction in that region. This fossil offers critical insights into the group's distribution and history.

半翅目／ドミニカ琥珀に入ったツノゼミの化石（左）Centrotinae gen. sp. in Dominican amber (left) (BL=6 mm); 現生の近似 / 近縁種である台湾産のザウターイカリツノゼミ（右）*Leptobelus sauteri*, extant species (right) (BL=9 mm) (Hemiptera: Membracidae: Centrotinae).

水面を走る

アメンボ Water strider

アメンボと聞いて思い浮かべるのは水に浮くことだが、その原理は比較的簡単である。脚の先（跗節）に毛が密に生えており、それが油を含んでいる。毛の間にたまった空気が層となって水を弾き、水面の表面張力の上で、沈まずにいられるのである。本種は渓流や細流に生息し、前後左右に素早く水面を走り回って、流れてくる餌を探す。普通種ではあるが、拡大してみると思わぬ発見があった。跗節の先端に虹色の鱗毛が規則正しく並んでいたのである。多くのアメンボでは毛が密に生えているだけであるが、本種はその点でとても変わっている。ただ、これにいったいどんな機能があるのかは不明である。

The principle of floating on water by water striders is simple. The ends of their legs (tarsi) are covered in dense, oil-coated hairs. These hairs trap air, forming a layer that repels water, allowing them to stay afloat due to surface tension. The tarsi of this species are also covered with iridescent scales arranged in order. While most water striders just have dense hairs, this feature is unique, though its function remains unknown.

半翅目／シマアメンボ *Metrocoris histrio* (Hemiptera: Gerridae: Halobatinae) (Japan). BL=5.4 mm.

土の色、空の色

アゲハチョウ Swallowtail butterflies

アフリカを代表する巨大なアゲハチョウ２種である。日本のアゲハと同属であるが、雰囲気はまったく異なる。世界的に見ても、非常に特異なアゲハチョウである。どちらもアフリカ中央部から西部の熱帯林に生息し、同じ場所で見つかることもある。２種の色彩は対称的で、ドルーリーオオアゲハは赤と黒を基調とするアフリカの昆虫に多い独特の色彩で、派手ではない。いっぽう、ザルモクシスオオアゲハは非常に鮮やかで、吸い込まれるような青色である。驚くことに、どちらも幼虫が見つかっていない。

These are two of Africa's most giant and most distinctive swallowtail butterflies. Both species live in the tropical forests of Central and Western Africa and are sometimes found in the same locations. Their colors contrast: one has a subdued red and black pattern, while one is strikingly vibrant with a mesmerizing blue. Interestingly, the larvae of both species have yet to be discovered.

鱗翅目／ドルーリーオオアゲハ（左） Drury's Swallowtail, *Papilio antimachus* (left); **ザルモクシスオオアゲハ（右）** Zalmoxis Swallowtail, *P. zalmoxis* (right) (Lepidoptera: Papilionidae) (C. Africa). **原寸** Actual size.

溝と点刻

エンマムシ Clown beetle

大きい、珍しい、かっこいいというのは、人気昆虫の重要な条件である。本種は日本産種のなかではその筆頭にあげられる。大きいに関しては、「エンマムシ科の中で」という注釈を入れなければならないが、珍しいのは間違いのない事実だし、かっこいいに関しても異論は認めない。体長は8ミリメートルほどで、エンマムシとしてはかなりの大型種でありながら、これまでに日本で見つかっている産地は、おそらく10ヵ所にも満たない珍種であり、縁取りに溝のある前胸や不規則な点刻など、なにより形態が素晴らしい。クサアリのなかまの巣に生息し、アリを食べる様子が観察されていることからこの名がある。

Size, appearance, and rarity are key traits for popular insects, and this species is a top example in Japan. While "large" here applies within its family, Histeridae, it is certainly rare and striking in appearance. At about 8 mm long, it is large for a clown beetle and has been found in fewer than ten locations in Japan. Its grooved, bordered thorax adds to its appeal. This species was named after one of the authors.

鞘翅目／アリクイエンマムシ *Margarinotus maruyamai* (Coleoptera: Histeridae: Histerinae) (Japan). BL=8.4 mm.

微毛は命綱

マツモムシ Backswimmer

マツモムシは半翅目（カメムシ目）の水生昆虫で、水面に落ちた昆虫や水中の小さな生き物をおそい、ストロー状の口で刺して捕食する。腹部を上に向けて長い後脚でボートを漕ぐように泳ぐが、そのとき、体の表面が銀色に光って見える。それは、体の表面が水をはじく微毛で覆われており、薄い空気の層を作っているためである。この構造をプラストロンという。水中の酸素がそこに溶け込み、逆に二酸化炭素を溶け込ませる（放出する）ことができるので、あまり頻繁に息継ぎをすることなく、水中で長い時間活動することができるのだ。体表を拡大すると、マット状の微毛の上に黄色い剛毛が生えている。剛毛の基部には感覚器があり、空気層の状態を把握するのに役立っている。

半翅目／マツモムシ *Notonecta triguttata* (Hemiptera: Notonectidae) (Japan). BL=11 mm.

Backswimmers are aquatic bugs that prey on insects and small creatures by stabbing them with its straw-like mouth. It swims on its back, rowing with its long legs, and appears silver due to a thin layer of air trapped by fine hairs on its body, called a plastron. This air layer allows the bug to extract dissolved oxygen from the water and release carbon dioxide, enabling it to stay underwater for long periods without needing to surface for air. Under magnification, the body surface reveals yellow setae on the plastron. Sensory organs at the base of each setae help the bug monitor the condition of its air layer.

極彩色とりどり

アフリカのカミキリ African longhorn beetles

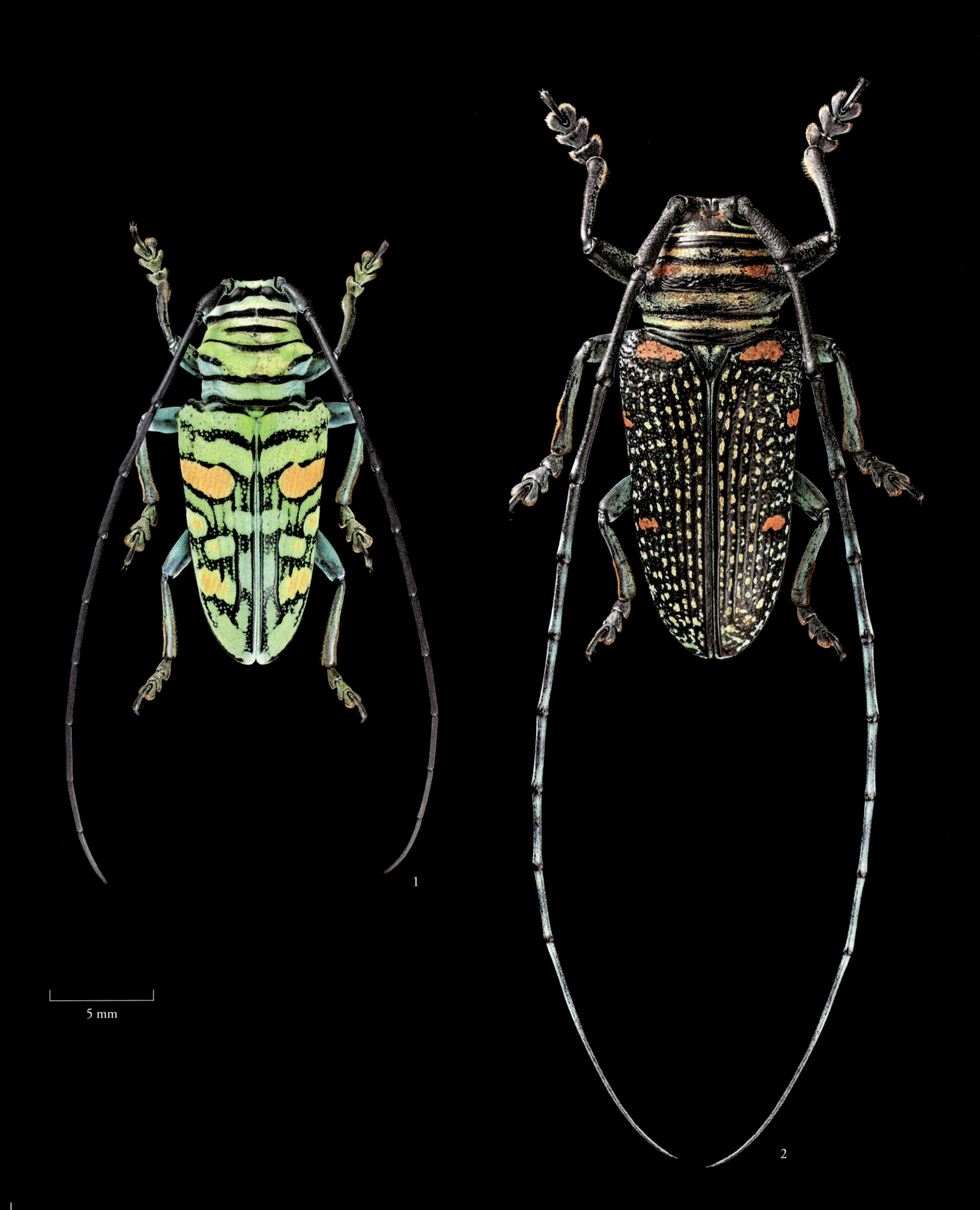

ホウセキカミキリはアフリカを代表するカミキリムシの一群である。カミキリムシのなかでも進化した、フトカミキリ亜科というなかまに属する。オスは立派な大顎をもち、それを用いてメスをめぐって戦う。その名の通り大変美しい模様をもつが、これは色の異なる細かい毛が集まって織りなされている。それぞれの毛はフォトニック結晶という微細構造をもっており、特定の波長の光を反射することで構造色を発する。水分や色素も多少関係しており、このなかまは標本にすると色あせてしまう。撮影した個体は非常に新鮮で、自然のままの色を表している。橙色や緑のデザインはアフリカのファッションを彷彿とさせる。

The tribe Sternotomini is a group of jewel longhorn beetles from Africa. Males have large mandibles used in fights over females. As their name suggests, these beetles are adorned with beautiful patterns created by tiny hairs of different colors. Each hair has a photonic crystal structure reflecting specific light wavelengths, producing iridescent colors. The freshly photographed individuals retain their natural brilliance, with orange and green patterns reminiscent of African fashion.

鞘翅目／ホウセキカミキリのなかま (Coleoptera: Cerambycidae: Sternotomini). 1, *Sternotomis amabilis*; 2, *Zographus regalis*; 3, *S. rufomaculata*; 4, *S. pulchra* (C. Africa).

あやとり

シロアリモドキ Webspinners

糸を出す昆虫というのは目レベルではあまり多くない。シロアリモドキは数少ない糸を出す昆虫の一群である。聞いたことのない人が多いと思われるが、紡脚目（シロアリモドキ目）という小さなグループに属し、日本には温暖な地方に数種が産するのみで、滅多に人目に触れることはない。糸を出す部分がまた面白い。脚の先のほうが丸くふくらんでおり、そこで絹糸の成分を生産する。そして、そのわきに毛のような突起が並んでおり、その先から糸を出すのである。この糸で木の幹や岩の表面などに巣をつくり、その中で生活している。

A few insect groups produce silk, and Embioptera (webspinners) is one of them. What makes them fascinating is the unique way they do so. The ends of their legs are rounded and produce silk components, with tiny hair-like projections nearby from which the silk is spun. They use this silk to create nests on tree trunks or rocks where they live.

紡脚目／タイワンシロアリモドキ *Oligotoma humbertiana* (Embioptera: Oligotomidae) (Japan). BL=5.7 mm.

5 mm

利那のメタルボタン

ジンガサハムシ Tortoise beetles

このなかまは日本にも数種が生息し、どれも生きているときは美しい。この「生きているとき」というのが重要で、この虫の場合は、死んでしまうと光沢が弱くなるどころではなく、真っ茶色の地味な甲虫に変わってしまう。体内の水分が色彩に関係していて、ケースの中に入れて、しばらく置いておいても、すぐに色が変わってしまうほどである。アフリカでは非常に種数が多く、写真も同じ場所のヒルガオ科の植物に見られたものである。生きているときは貴金属そのもののような色合いで、縁に透明な部分があるのも面白い。

鞘翅目／ジンガサハムシのなかま *Aspidimorpha* spp. (Coleoptera: Chrysomelidae: Cassidinae) (C. Africa).

The metallic colors of these tortoise beetles are stunning when alive. The key is "when alive" because their vibrant sheen fades completely once they die, leaving them as plain brown beetles. The beetle's color depends on its internal moisture; even when placed in a case, the color fades quickly. In Africa, there are many species of this group. These species in the photo were seen on some plants of the morning glory family in a single place.

原寸

箱に眠る記憶

シジミチョウ Lycaenid butterflies

日本ではさまざまなチョウが絶滅の危機にある。日本に生息するチョウの約 1/3 が危険な状況にあるともいわれている。ただし、これはチョウが多くの同好者によって調べられているからこそわかっているのであって、実は多くの昆虫が似たような状況にある可能性が高い。昆虫は生態系を、ひいては私たちの生活を支える重要な存在であり、その多様性の消失は深刻な問題である。これらのチョウは絶滅かそれに近い状態にあるとされる 2 種で、これら乾いた古い標本は環境問題を象徴するとともに、過去の生息の貴重な生き証人となっている。

鱗翅目／オガサワラシジミ（左）Ogasawara holly blue, *Celastrina ogasawaraensis* (left); ツシマウラボシシジミ（右）Tsushima blue Quaker, *Pithecops fulgens* (right)(Lepidoptera: Lycaenidae) (Japan).

Many butterfly species in Japan face extinction, with about one-third of the country's butterflies considered at risk. This fact is well-documented due to the efforts of many butterfly enthusiasts, but many other insect species may be in similar danger. Insects play a crucial role in ecosystems and, ultimately, in supporting human life, making their decline a severe issue. The two butterflies shown here are species considered near extinction, and these old, dried specimens serve as valuable evidence of their past existence.

原寸

カメルーン遠征のこと

法師人 響 *Hibiki Hoshito*

昆虫は死後に色が変わっていく。生時の色を完璧に残す虫は稀だ。そのため、美しい体色を楽しんでもらうなら、なるべく鮮度の良い標本を撮影したい。そのために、国内にとどまらず、海外に出かけて採集と撮影を行う。しかし、前作『驚異の標本箱』は製作中に疫禍が起こり、海外での撮影は叶わなかった。前作刊行後、丸山さんが「次は海外へ行きたい」と語ったことを覚えている。

そして2023年末に『神秘の標本箱』製作の打診があった。ちょうど同じ頃、丸山さんがカメルーンで行う調査の同行者を探していたので、手を挙げてみたところ、調査と並行して本書のための撮影を行うことになった。

2024年4月、韓国とエチオピアを経由し、カメルーンの首都ヤウンデへ到着した。そこから南に70キロメートルほど離れた村が今回の滞在地である。ここには丸山さんが信頼を寄せる採集名人がいる。見たい虫のことを伝えると、あっという間に探し出してしまうのだ。彼らの虫の採り方からは学ぶことが多かった。

印象的なのはホウセキカミキリ（144-145頁）の採集だ。森の中の伐採地で、ジョセフさんがホウセキカミキリを次々に見つけてくる。私も自分で見つけたいと思ったものの、なかなか見つからない。よく飛ぶうえに、好む樹種がわからないのだから当然だろう。しばらくの間、撮影のために綺麗な個体を探しながら、各種の生態を調べる日々が続いた。

オオキバドロバチ（54-55頁）も思い出深い。メスはよく見られるが、オスを探し出すのは難しい。虫採り名人たちの力を借りてなんとか撮影することができた。

さらに、森で見つけたジンガサハムシには驚かされた。綺麗な種だと喜んで捕まえたところ、数分で色が変わったのだ。死後に変色することはよく知られているが、生き

1　左手にザルモクシスオオアゲハ／2　伐採された材木には多くの虫が集まる／3　地衣類を食べるウメノキゴケツユムシ／4　微動台やカメラ内のフォーカスブラケット機能を使ってピントを移動させる／5　滞在したロッジ／6　虫採り名人のジョセフさん（左）とフダさん（右）／7　夜は明りに集まる虫を探す。昼も夜も休み無しだ／8　写真を確認する丸山さん。納得のいくまで一種の虫を何度も撮影し直した／9　ホウセキカミキリを採って喜ぶ筆者／10　停電すると発電機をつけてくれるが、ガソリンが無くなると朝まで真っ暗だ／11　ホウセキカミキリのオスは大顎が目立つ／12　ロッジから見た夕焼け／13　森の中の小道は村人の生活道路にもなっている／14　オオキバドロバチのオスと巣／15　素早い網さばき！

ているうちに色が変わるなんて！ まるでカメレオンのようだ。丸山さんが生きた個体を撮影しているので、美しい質感を楽しんでほしい。(148-149頁)

極めつけはウメノキゴケツユムシ（70-71頁）だ。枝の上にいるところを撮影していると、その場の地衣類を食べ始めた。普段は地衣類に身を潜めながら、それを食べて暮らしているのだろう。形態と生態を、相互に補完しながら作り出す擬態の妙を感じる瞬間であった。

そして、この地にはアフリカを代表する蝶であるザルモクシスオオアゲハ（139頁）がいるという。これを狙い、丸山さんと川べりへ出かけた。すると前を歩いていた丸山さんが叫ぶ。

「ザルモクシスだ！」

網が振られ、あっという間にザルモクシスオオアゲハはその手中へと収まった。その時、調査と撮影で疲れていた丸山さんが少年のような優しい微笑みを見せた。この笑顔はどんなに希少な虫でも価値の及ばない、大切なもののように思えた。

撮影については日本とは異なる方法をとった。特に精密な電子機器が壊れることは避けたい。そこで、システムの要である自動撮影用のレールを排し、簡易的な撮影セットを用意した。

さらに、頻発する停電には苦労させられた。涼しくなった夜に撮影を進めたいのだが、ほぼ毎日停電が起きてしまう。時間をかけて標本を作り、いざ撮影！という時にこれでは泣きたくなる。

こうして苦労は多かったものの、初めての西アフリカということもあり、見るもの全てが新鮮で楽しい遠征となった。同行してくれた仲間たちには、この場を借りてお礼を申し上げたい。

オオトラカミキリ

Xylotrechus villioni (Coleoptera: Cerambycidae) (Japan). BL=32.5 mm.

世界的に見ても最大級のトラカミキリで、低標高地では晩夏を飾る最後の大物と呼ばれる。9月初旬ごろの猛暑かつ晴天の日、モミの低い位置に、産卵のために降りてくるメスが採集される。このように採集に関わる重要な生態が解き明かされた今でも採集難度は高い。撮影したのは大型個体で、前胸と比べた際の前翅の太さが印象的。

In Japan, this is the last notable longhorn beetle of late summer in lowland areas. It has a striking tiger-like pattern that resembles a hornet. On hot, sunny days, females can be found descending to low fir trunks to lay eggs. While this crucial behavior is now understood, capturing this beetle remains challenging.

あとがきにかえて——1

二番目に好きな虫

吉田 攻一郎
Koichiro Yoshida

「二番目に好きな虫」である。一番好きな虫であれば、ただのひとつを挙げるのにそれほど悩まない。しかし、二番目となると少し悩ましい。あれもこれもと浮かび上がる。複数でよいとなれば、しばらく語れるくらい好きな虫は明確にある。けれど、二番目をひとつ、となると悩むのだ。そうして本書の撮影を進めながら考えて、答えが出た。この虫である。

選んだ根拠は、初採集時の感動の大きさと、その後も挑戦し続けているかというふたつの点である。そう。初採集の光景はいまでも鮮明に思い出すことができる。誰もいない、自分だけのモミの森の中で、急斜面の上り下りを繰り返し。蒸し暑さとセミの声。条件は聞いている通りでよさそうなのに、さすがに難敵だ。そろそろあきらめ移動するかと思ったその時に、斜面の上の方にそれまで見落としていた一本のモミが見えたのだった。一瞬考え、木の根草の根を掴み、斜面をよじ登った。近づいて見てみると、それは直径20センチメートル程度の若い木であった。根元には白く輝くヤニしぶき。体勢を整え、視線を上げていく。瞬間、時間が止まった。その光景を思い描いていたはずなのに、現実にすべてを吹き飛ばされた気がした。

私は「威風堂々」を感じる光景に弱いらしい。モミの幹、頭上1メートルくらいのところから一歩、また一歩と降りてくるその歩みは、まさに威風堂々。森の王者がそこにいた。この虫の出現する時期と環境では、他に狙う虫があまりない。そんなことから夏の最後の運試しといった趣もある。幸い、その後にも何度か採集の幸運に恵まれたものの、以降この10年ほどは採集できていない。それでも「今年こそは採れるのではないか」と毎年最低一回は挑戦してきた。あの時の光景をもう一度見てみたい。脳内を満たす恍惚を求める行動を、どれほどその虫が好きなのかの判断材料とすることにきっと異論はなかろう。

なぜ今回、あとがきのお題が「二番目に好きな虫」になったのかというと、前作『驚異の標本箱』に著者それぞれの一番好きな虫が入っているからである。私の一番好きな虫についても、思い返してみれば「威風堂々」だ。青空に突然現れ、威風堂々と旋回飛行をする。空の王者である。昆虫のどういうところに惚れ込むか、ということになると、どうやら私はその昆虫の初登場時の姿に魅力を覚えるようだ。

さて、本書が世に出る2024年はどうだったのかというと、撮影の追い込みと標本の準備を最優先としたため、一番目、二番目に好きな虫に、初めてまったく挑戦しない年となった。『驚異の標本箱』から4年。当時よりも深度合成撮影を手掛ける方も増えていることと思う。ただきれいに撮影するだけでは不十分と、昆虫の神秘をどのように伝えられるか、試行錯誤の集大成となった。好きな虫への挑戦を返上して取り組んだ本書、いかがだったであろうか。本書を手に取り、昆虫の神秘を、図鑑と異なる切り口から楽しんでいただけたとしたら望外の喜びである。

撮影担当ページ
P4-5, 10-11, 18-21, 32-35, 44-45, 48-49, 60-63, 66-67, 72-75, 82-83, 86-87, 90-91, 94-95, 102-105, 116-117, 122-125, 128-131, 136-137, 140-141

オオヨツボシゴミムシ

Dischissus mirandus (Coleoptera: Carabidae) (Japan). BL=16 mm.

河川敷や湿地に生息し、ナメクジだけを食べる。冬期は朽ち木の中で越冬する。刺激を与えると捕食者が忌避する液体（*meta-* クレゾール）を噴射するが、私はこの香りが大好きだ。

This beetle species lives in riversides and wetlands, feeding exclusively on slugs. During winter, it hibernates in rotting wood. It sprays a repellent liquid, *m*-cresol, when threatened to deter predators, though I find the scent delightful.

あとがきにかえて——2

二番目に好きな虫

法師人響

Hibiki Hoshito

前作『驚異の標本箱』は駆け出しの写真家であった自分にとっては指折りの大きな仕事だった。頑張りの甲斐あって、反響は大きく、2024年現在でも「読みましたよ！」とお声かけいただくことが多い。今回はその第二作をつくるにあたって、『驚異の標本箱』の製作にお誘いいただいた丸山さんと吉田さんのお二人に、恩返しができたら、という思いがあった。ところが、いざ撮影を始めてみると、お二人が明らかにパワーアップしている！ 恩返しどころではない！ その勢いに振り落とされないよう、必死に写真を撮ることになった。

印象深いのは、チョウトンボの撮影のことだ。実は前作でも撮影を試みていたのだが、思うように撮れず、掲載を断念していた。だが、今回はイメージしていた以上の画をつくることができた。前作の頃の自分と比較して、成長を感じることのできた瞬間であった。『神秘の標本箱』において、私が撮影した写真の中では出色の出来となったと思う。

また、私は著者陣の一人であるが、前作のあとがきに書いた通り、丸山さんと吉田さんのファンでもある。お二人の新作の写真が見られることがとても嬉しかった。同時に学ぶことも多く、製作期間はあっという間に過ぎていった。その間、撮影への協力や励ましの言葉をいただいた皆さまには、この場を借りて厚くお礼を申し上げたい。

さて、「標本箱シリーズ」第二作ということにちなみ、あとがきでは「私の二番目に好きな虫」と銘打って、それぞれお気に入りの虫を紹介することになっていた。近年は特定の種にこだわらず昆虫の「箱推し」となりつつある自分としては大変悩ましいテーマだ。そこで、これを「それなりにどこにでもいる馴染み深い虫」と解釈し、私からはオオヨツボシゴミムシ*Dischissus mirandus*を紹介したい。ゴミムシとしては比較的大きな体躯に、鮮やかな黄色い紋が目をひく。本種は、私が初めて標本にしたいと思い採集した虫である。ラベルには2014年11月23日とあり、ちょうど10年前のことであった。希少というわけではないが、それなりに生息する環境を選ぶ虫という印象があり、今でも見つける度にじんわりと嬉しくなる。私にとっては、年に数度のやりとりがある旧友のような存在だ。

撮影担当ページ
P6-7, 12-13, 22-23, 28-29, 36-39, 46-47, 54-57, 68-71, 78-79, 92-93(幼虫), 98-99, 106-115, 142-147

ミミカホウセキゾウムシ *Eupholus mimikanus* と
スハンダホウセキゾウムシ *E. suhandai*
(Coleoptera: Curculionidae) (Oceania) . Both BL=25 mm.

解説は 16 頁を参照。
See, p.16 for explanation.

あとがきにかえて――3

二番目に好きな虫

丸山宗利
Munetoshi Maruyama

まず好きというのが難しい。私にとってそんな昆虫はあまりにも多いからだ。そして何をもって好きというのかも難しい。もっとも原初的な感覚となりうるのは外見だろう。美しいとかカッコいいとか、見た目を好ましく思える昆虫は当然好きになる。世界に美麗な昆虫はあまりに多く、その点で選択肢はほとんど無限である。また、思い出深さも重要といえる。子供の頃に飼育していたとか、はじめて捕まえてびっくりしたとか、記憶の奥底にあるなんらかの引っ掛かりを与えてくれることも好きの要素となる。逆に、知識のみで接点がなく、強いあこがれをもつこともあるだろう。

さらに言えば、何をもって好ましい外見なのかと思えるかも難しい問題である。さきほどの思い出や憧れもその要素になりうるし、それらがなければ好ましく感じられない人も少なくない。また、さまざまな虫を知り、虫に詳しくなってくると、私たち虫屋の性として、珍しいというのもの重要な要素になってくる。これは非常にいやらしい価値観で、知識によって逆に感覚が狂ってしまうという悲しい性でもある。たとえばそのへんに飛んでいるアゲハチョウは美しいと思えない。しかし、そんなアゲハチョウでも、珍しい異常型であれば、なぜか飛びぬけて美しく見えてしまい、その良さを偉そうに友人に語ったりしてしまう。もちろん、そんなものも、虫に詳しくない人にとっては、多くは違いさえわからない程度のものである。

そんなこんなで私には好きな虫が無数にいて、恥ずかしながらとくに珍しいものに惹かれてしまう。しかし、できるだけ虫屋的な邪念を取り払い、心の中にあるかもしれない水晶玉のようなものに向かい合って、何が好きかを考えてみた。その結果、強いてあげるならば、ホウセキゾウムシのなかまとなった。残念ながら邪念はぬぐい切れず、ここでは珍しいものを選んで撮影してしまったが、標本箱の見開きのページにあるように、普通種も珍品も、どれもこれも美しい種ばかりである。こういうときに細かい説明はいらない。皆様の素直な心の目でこのゾウムシを鑑賞いただきたい。

第一弾の『驚異の標本箱』の出版はそもそもそれ自体が私たちにとって驚きで、順調に部数を伸ばしたことはさらなる驚きだった。そして今回の第二弾のお話をいただいた。無限の多様性を誇る昆虫といえども、被写体が入手でき、なおかつ面白い話ができる昆虫には限りがある。正直言ってかなり苦労したが、第一弾に劣らぬ美しい本ができたと思う。本書の出版を支えてくださった編集の小出真由子さんと小荒井孝典さん、デザインの鷹觜麻衣子さん、そして撮影にご協力くださった皆様には厚くお礼申し上げます。

撮影担当ページ
P8-9, 14-17, 24-27, 30-31, 40-43, 50-53, 58-59, 64-65, 76-77, 80-81, 84-85, 88-89, 92-93(成虫), 96-97, 100-101, 118-121, 126-127, 132-135, 138-139, 148-151

丸山宗利（まるやま　むねとし）

九州大学総合研究博物館准教授。1974年生まれ。東京都出身。昆虫学者。北海道大学大学院農学研究科博士課程修了。博士(農学)。国立科学博物館、フィールド自然史博物館（シカゴ）研究員を経て2008年より九州大学総合研究博物館助教、2017年より准教授。アリと共生する昆虫を専門とし、国内外で数々の新種を発見。より簡易な深度合成写真撮影法を考案し、研究のかたわら、さまざまな昆虫の撮影を行う。2020年にシリーズ前作となる『驚異の標本箱—昆虫—』を刊行し、2021年には『角川の集める図鑑 GET！昆虫』で総監修をつとめた。（ともにKADOKAWA）。ほか監著書多数。2024年日本動物学会「動物学教育賞」を受賞。

Xアカウント　丸山宗利 Maruyama @dantyutei

吉田攻一郎（よしだ　こういちろう）

工業デザイナー。1976 生まれ。山梨県出身。幼少期より昆虫、植物、魚類、鉱物など自然に関するものに興味をもち育つも、中高大と自然と無縁の生活を送る。社会人になり、子供のころ図鑑で憧れた虫を実際に手に入れられると知り、再燃してのめりこむ。その後、採集・撮影も行なうようになり「ムシトリアミとボク」というブログ（現在は更新停止中）で多くの昆虫好きとの交流が生まれた。昆虫、特に甲虫の、生態と連携した機能を備えた構造や、色彩・造形に魅力を感じている。いつまでも変わった虫を集めて手に取って眺めたいと思っている。2020年にシリーズ前作となる『驚異の標本箱—昆虫—』を刊行。2021年には『角川の集める図鑑 GET！昆虫』で標本協力。（ともにKADOKAWA）。

Xアカウント　かりめろ《吉田攻一郎》@KOH16

法師人響（ほうしと　ひびき）

昆虫写真家。日本自然科学写真協会会員。1999年生まれ。茨城県出身。幼少のころから昆虫に親しみ、2017年に写真の撮影を主体とするようになる。現在はネイチャー系クリエイター集団「Tokyo bug boys」として活動中。世界中でさまざまな野生生物を撮影する傍ら、故郷茨城で昆虫の分布調査も行う。日々新しい表現に挑んでおり、生物の多様性や自然の美しさを多くの人に届けたいという思いが活動の源になっている。2020年にシリーズ前作となる『驚異の標本箱—昆虫—』を刊行し、2021年には『角川の集める図鑑 GET！昆虫』で昆虫撮影を担当（ともにKADOKAWA）。ほか多数のメディアへ活動の場を広げ、現在『所さんの目がテン！』（日本テレビ）に出演中。

Xアカウント　法師人響 昆虫写真家@tritomini

デザイン／鷹觜麻衣子
校正／亀澤洋
謝辞／青井光太郎、安西稔、安藤真人、大塚昌平、小川浩太、小野広樹、柿添翔太郎、金子みずき、鳥山邦夫、草野勇茉、阪本優介、佐藤仁、白川真理、城戸克弥、関峻大、菅原大輝、瀬筒秀樹、田代巧、田中久稔、鳥羽明彦、長島聖大、中瀬悠太、名嘉猛留、西田恒介、畑祥雄、花谷達郎、林紗南美、土方俊雄（故人）、日野真人、平井文彦、福菌貴史、藤井告、前田健太、吉富博之、Itsarapong Voraphab、Sasitorn Hasin、愛媛大学ミュージアム、九州大学遺伝子資源開発研究センター、九州大学総合研究博物館、むしとりチャンネル
カバー写真／吉田攻一郎（掲載種はベニスカシジャノメ *Cithaerias pireta*）

神秘（しんぴ）の標本箱（ひょうほんばこ）　－昆虫（こんちゅう）－

2024年12月11日　初版発行

著　者　丸山 宗利、吉田 攻一郎、法師人 響
発行者　山下直久
発行　株式会社KADOKAWA
〒102-8177東京都千代田区富士見2-13-3
電話0570-002-301（ナビダイヤル）
印刷・製本　TOPPANクロレ株式会社

●お問い合わせ
https://www.kadokawa.co.jp/（「お問い合わせ」へお進みください）
※内容によっては、お答えできない場合があります。
※サポートは日本国内のみとさせていただきます。※Japanese text only

定価はカバーに表示してあります。

ISBN978-4-04-115134-1　C0645